The Journey Home

Michael Yanuck MD PhD

ISBN: 978-1-946600-31-8

DEDICATION

In loving memory of
Sal, Pat and Michael...
.

Names of people and places have been changed, and although not always presented in the order they occurred, the events of this story are otherwise based on real events.

Part One

During the summer that I returned to college to advance the work that would springboard the development of a cancer vaccine, my life was irrevocably changed when a relationship with a beloved friend flowered and bloomed…

I was nothing, no one, I was everything to her, I was hers.
~Sharon Olds, The Wellspring...

INTRODUCTION CHAPTERS 1 – 30 (ABRIDGED)

July, 1984

Around noon I heard the lab door open behind me, followed by the sound of gentle footsteps, then arms that descended about me.

"Whatcha doin'?" she whispered.

"Not much," I said, every fiber of my being relaxing under her touch. "Gonna be a long day."

"Ready for lunch?"

"Let's go…"

Overhead the skies were clear and blue as we rode our bikes into the fields beyond the Physics Building. In the shade of a tree, we ate from the contents of her basket and shared the stories of our lives.

"When I was a boy," I began, "there was a family that lived down the block. They were a bad bunch. Bullies, every one of them. The Gilmores. One day the youngest boy, who was still probably 10 years older than me, stood in the middle of the road, and as I passed him riding on my bike, he did this windmill punch that struck me square in the back. I felt the wind knocked out of me, and I saw this white light. My back hasn't been right since."

I broke off.

I was nothing to him, I thought. Just target practice. But he changed everything for me. Leaving me with an aching back and a crooked spine for the rest of my life. I could still feel his fist in my back.

Then, I felt her fingers.

"Is this where it is?" she asked. "I remember us working on it a year ago."

Yes, that was the night in the dorms when she and some of the other girls came and offered massages to some of us guys.

"And we were trying to get it to go into place," she continued, "but it wouldn't."

Yeah, I thought. Stuck.

"Back then, it was just a little bit out of place – like only a few millimeters," she added. "Now, it feels more like a couple of centimeters."

I nodded, grimly.

Worse, I thought.

But sinking into her touch, I experienced a feeling of relief and sense of kinship – as though she were a sister from another mother.

Then, more memories of the events of that day surfaced again.

"When I got home," I continued, "my father was working in the garage. I was crying as I told him what happened. 'Go hit him back!' he says. But there was no 'hitting him back' – That Gilmore boy was easily twice my size.

"So, my father goes storming off, and when I looked around the garage, I saw him walking straight towards the Gilmore boy, still standing there in the middle of the street. My father was a head shorter than that boy was. But when he gets to him, he grabs him by the neck, swings him around and marches him to his house, talking to him the whole way, in what I imagined was that mean way of his."

"And watching," I concluded, "I couldn't help but feel sorry for him."

Breaking off, I felt lost in a feeling of despair that I'd never come out of.

Then, Bubbles began to speak.

"When I was little," she said, "my father took our bikes away."

"What happened?" I asked, moved.

"Another girl in the neighborhood got hit by a drunk driver," she responded, matter-of-factly. "She was riding her bike, and when she died, he forbid us from ever riding our bikes again."

She broke off and looked away, but not before I saw a tear form in her eye.

Gazing at her, I thought about her father – Her description of sitting with him on a park bench; meeting him a year ago, and his telling me the importance of locking the door.

In my mind's eye, I knew that he had only acted to protect the daughters he loved back then; but I also recognized it made her sad.

Then, I was gripped by an odd and unfamiliar sensation: It was as though I were somehow outside myself – existing only in spirit – and all that mattered was my cherished friend.

I reached to her.

"I feel for you," I said. "I feel for you..."

Part Two

The year between college and medical school, I studied abroad at the Weizmann Institute of Science in Rehovot, Israel. There, I studied the newly discovered p53 oncogene, which would come to be recognized as the most commonly mutated gene in all of human cancer and serve as the target for the cancer vaccine I would develop years later.

In the interim, my beloved grandfather died and a deep-seated gloom set in.

Then, my father visited…

pity this busy monster, manunkind...

~E. E. Cummings...

CHAPTER THIRTY-ONE

January, 1987

Along a beachfront in Tel Aviv a vigorous breeze blew in from the sea. The sand shifted and the wind howled. My father donned a black, leather jacket, appearing in spirits and only mildly affected by the cold; I noticeably shuddered, the cold ripping through my sweater; muscles tight, body hunched, face long, eyes, hopeless.

"That's quite a Chi you're wearing," my father commented, gesturing to the choker around my neck. "Is that something you picked up here?"

"Grandpa gave it to me," I answered, trembling. "On the day he died, I went to the Wailing Wall and put it on. I hadn't worn it in years – Not since the summer with..."

"Oh, no!" he exclaimed, disgusted. "We're not going to talk about that again?!"

He hesitated.

"You'll get over her! Trust me!"

He walked ahead.

"A whole country full of beautiful girls and all you can do is mope about this one!"

He shook his head.

"Why don't you pick yourself up and get over her?!"

The sand blew coarse against our faces. My father turned and strode ahead. I watched his back as he ascended the stairs and walked away...

Part Three

At the National Institutes of Health, I parlayed my knowledge of the p53 oncogene into the development of a cancer vaccine that would ultimately revolutionize the treatment of cancer and usher in the science of Neoantigen Tumor Immunotherapy that would come to be used worldwide.

Not long after the vaccine received FDA-approval, though, my leg was injured and doctors told me I wouldn't walk normally again and could expect chronic pain for the rest of my life…

Those who have a 'why' to live can bear with almost any 'how'...
— Viktor E. Frankl, Man's Search for Meaning…

CHAPTER THIRTY-TWO

June, 1992

I have to go on, I thought, but I fear my life has changed. Perhaps, it was changed before, and I didn't recognize it. Instead, I pushed ahead, and this is the result – lying here, day after day, dreaming, wishing for the recovery that never comes.

My father comes tomorrow. Why does he have to come? I begged him to let me see this thing through alone. But, no. He insists. Even if it is a token visit, he wants to come look in on me. What harm could it do? he says. He would stay a couple hours, then be on his way.

He asked why I'd taken the receiver off the hook, saying my mother was calling him, day after day – saying she was worried, and couldn't get through?

She'd been anything but comforting.

"Why are you in so much pain?" she'd told me. "Take a Motrin – that should take the pain away. I don't understand why you're in so much pain..."

The hours passed. Nighttime was the worst. The stillness in the room. The lifelessness of things around me. I lay on the floor. Hard surfaces were all I could tolerate. Anything else set off a cascade of pain. Things were kept within an arm's distance, as reaching too far led to spasms.

I got onto the skeleton of the bed. Weeks ago I'd removed the mattress, and slept on the plywood undersurface. My prominences ached, still, I couldn't bear being on the floor anymore – something vulnerable about being so low to the ground – Like an animal, peering up all the time.

I can't go on lying here, I thought, day after day. I have things to do – plans to make.

I sat at the desk and began to write. Within moments, though, the pain began – a throbbing in my hip, like being in the grip of a metallic claw closing around me, and the muscles stretched so taut that, at any moment, they'd either snap, or collapse into steel.

I reached for the phone, and dialed my father. To hell with superficialities. To hell with token visits. To hell!

I hung up the receiver.

No, I thought. Rash decisions – outbursts of anger – that seems all I'm able of do. Perhaps, I need help? Perhaps, she can help me?

I dialed her number.

"May I speak to Bubbles?"

"This is she."

"Hi Bubbles. It's Mike."

"Hi Mike!" she gushed. "I recognized your voice."

"I recognize yours," I said. "How are you?"

"Alright. How about you?"

"I've been better," I said.

"Oh," changing her tone. "What's wrong?"

"I've done something to my leg," I said. "I haven't been able to walk for a while."

"How long?" she said.

"About four months."

"Well, you beat me," she said. "I've had pneumonia for only two."

I hesitated. She began to joke.

"Hmm, well," she said. "Nice talking to you. I guess I'll be seeing you."

"Sure. Talk to you later," I chided. "No, really Bubbles. Are you alright?"

"I'm getting better," she said. "For a while I was really sick."

"Fever? Shortness of breath?"

"Oh, yeah, all that," she said. "I could hardly breath."

"Did they find the cause?"

"No," she said. "They're treating me with antibiotics."

"Were they able to culture anything?"

"No. I still have a few white spots on my X-ray, but they're getting smaller."

"Oh, I'm glad," I said and lowered my gaze.

"What's wrong with your leg?"

I answered, detached.

"Do you know how it happened?" she said.

"It started while I was playing basketball with my uncle. It felt tight, so he offered to help by stretching out the muscles. He told me to lay on the ground, then applied his full body weight against the leg. Since then, it's just got worse."

"Have you seen a doctor?" she said.

"Yes, many," I said. "They can't seem to diagnosis it. One sports doctor says it's a torn hamstring; another claims it isn't. An orthopedist says a pinched nerve; a neurologist says neuropathy. They've done all manner of tests – X-rays, MRIs. Everything comes back negative. I don't think they know what they're dealing with, and I'm just becoming another one of those chronic pain patients they can't find an answer for and can't stand."

She hesitated.

"I think that's hard," she said. "It's hard when they can't give you an answer. Do you think it could be stress related? It just seems like you're always doing something so important. It all seems so stressful. You're always under stress."

"Maybe," I said. "You know, Bubbles, when I came to the National Institutes of Health, I wanted to do something for the thousands – for the millions. But, now, when I'm alone in my room – and it's dark – it's not the thousands that I see – not the millions that I think about – but, rather, it's the moments – moments in my life that I think of."

"It's normal to think that way," she said. "I think everybody does, especially when you're not feeling well."

"Yes, but I wonder. I wonder if I'm not doing the right thing anymore?..."

I broke off.

"Mike, I'm sure whatever you do, people will benefit," she said.

She hesitated.

"Does anyone from your family call?"

"My mother calls every night," I said. "She asks me why Motrin won't help?"

"And you brother?"

"Never," I said. "Not once."

"Your friend, Ben?"

"He calls, but he has his own problems," I said. "His marriage isn't going well."

I was quiet.

"My father wants to come out here," I said.

"Well, isn't that good?" she asked.

"I asked him not to."

"Why?" she said. "All this time it seems like you've been asking for some attention."

"Bubbles, you know how I feel about my father."

"Well – yeah – but if he's coming all that way just to see you."

"Actually, he's just staying a few hours on his way to New York."

"Oh," she said. "Then it's a stop-over."

"That's right," and in the pause that followed, I experienced a feeling of emptiness in my chest.

"Do you have a girlfriend?" she asked. "Or someone there with you?"

"I guess I did meet someone," I responded. "Her name was Michal. She was Israeli and quite simply the best dancer I'd ever seen. She whirled around like a blond fury. I remember watching her, thinking, perhaps, one day, if I get really good at this, I'll ask her to dance. It was on that very night that she asked me.

"But, as we began dancing more, I found myself wanting to be with her. It had been so pure in the beginning – all I'd wanted was to dance. Now, it was different – I was coming to be with her. And she was married.

"One evening I'd showed up at the dancing after a really disappointing day. My head was pounding, and there was pain in my leg. And, worse of all, I was kinda wondering what the hell I was doing there? I think that she could sense my confusion and stayed away at first. Then, she made an effort to cheer me up and asked me to dance. But I couldn't shake my feelings and wouldn't put on a brave smile on my face. I suppose – just once – I wanted her to see what I was feeling. I decided that I ought to leave. But she met me at the door, saying, 'I hope you feel better.' And when she went to kiss me, I pulled away..."

"And then one thing kinda led to another," she interrupted.

"No, nothing like that," I said.

"That was it?"

"Pissed her off, Bubbles," I responded. "How would you like it?... You go to comfort someone and they pull away from you like that?"

"Well, can you give me a number to give you a check-up call?" she asked.

I'd said too much, I realized.

"You take care," she said...

CHAPTER THIRTY-THREE

Midnight, and the pain was making its lunar appearance. Chip, chip, chip – like the workings of a stone carver's chisel, meeting its mark time and again till giving way to a new pain – this time, feeling like icicles drawn by slender threads, making their way down the length of my leg, while other took a different path.

In my mind's eye I saw fingers, tenderly, lovingly gracing the keys of a piano, and the faint outlines of her features, serene and pensive, as she studied the sheet music in front of her.

It was our summer, and she was dressed in a tank-top and shorts. Eastern art and parables of love decorated the walls of her sister's mobile home on the outskirts of the city. We sipped at freshly prepared strawberry daiquiris, and, in the ballroom, played pool. Having toured San Francisco during the day, we returned in the evening. Relaxed on the couch she held the toy terrier, as I ran my fingers through her hair...

"God didn't make you to work hard, Michael," she'd told me before parting. "God made you to love…"

CHAPTER THIRTY-FOUR

Opening the door my father was sitting with his legs up on my desk. The security guard had let him in. I'd gone to the metro station to meet him, but he'd walked past without noticing me.

I limped to the bed, and lay motionless on its mattress-less surface.

"I have a question for you," he said. "You remember that brown down-jacket I used to have... You know where it is?"

"I don't know where it is, Dad."

"I'll check with your brother," he said. "Maybe it's at your mom's."

He settled back into the chair.

"You know she's been calling me every day," he continued. "'What are we going to do about Michael?' she says. 'We have to help him.' She's been bugging the shit out of me! You know that!"

"Don't blame me if you're the one great love of her life," I said.

A smile flashed across his features. I turned on the bed.

"So, is it as bad as you thought?" he said. "Having your old man here?"

I looked at him.

"Dad, I really didn't want you to come here. I remember when grandma was dying of cancer. I wanted to see her – What did you say?... 'Leave her alone. What do you need to see her for? Write her a letter. What is it with you and face-to-face meetings?' My grandmother died in agony, and I never saw her. Now, you come and demand this 'face-to-face' meeting. I wonder to myself, 'What am I? Some catharsis, supposed to absolve you of your sins!'…"

"Michael, I have no regrets over my actions. No regrets. No reservations. If I had it to do all over again, I wouldn't change a thing. That's long settled, and I have no regrets for what I've done." His gaze always directed elsewhere. "But, I'll tell you something, Michael – when you get angry – it's scary.. The look on your face – the grimaces. You really hate. I certainly wouldn't want to be on your shit list."

"I wonder where I get it from," I said under my breath.

"I don't know," he said. "But it's scary. I'm serious."

"Dad, don't you remember? Conversation with my mother. You hung up the phone, and then turned on me. 'I'll kill her!" you said. 'I'll kill that bitch! I'll take away her house, and then I'll throw her out on the street! I'll kill her!'"

He broke into nervous laughter.

"I said that?... No?... Really?... I said that?..."

He smiled and thought, until a feeling of ease returned to his features.

"Still, I'll never understand," he said. "A person with as many accomplishments as you have – who's been to as many places – all the things you've been able to do. I'll never understand – this drive of yours – this intensity. What's the source of it? What's eating you?"

My face darkened, as I felt my features drawn inwards, my eyes pulled deep into their sockets.

"Because what I need I can't get!" I said. "I need love! And I can't find it!"

"But, Michael, all the people you've been with..."

"Nothing is enough! I'm never satisfied. Nothing is enough."

Breaking off, I lowered my head and closed my eyes.

They're not her.

And in the silence that followed, I felt alone in the world, and resigned to a place where no one could reach me, and none pull me back.

"Michael, I understand."

Moved and surprised, I looked to him.

"You remember Nikki?" he said. "You know how long we went out. There was a time one morning – we were sitting down at the breakfast table – when she asked me why we didn't just get married.

"I guess I had other things on my mind – other things I wanted.

"'I don't think it's such a good idea,' I said.

"Well, as you might expect, things kind of went down-hill after that. We split not long after."

He sat up.

"I'll tell you, Michael, I'll see a red-head now – one with Nikki's general physical appearance and features – and, I swear to God, I know it makes no sense – I know it's completely ridiculous – but I'll have to see her face. I'll, literally, have to physically get up close to her, and look at her. She could be in a passing car. She could be waiting in line at the market – I'll have to get up close to her, and look at her face, and be sure it's not her."

"I buried that woman," he continued. "I put her in the ground. I covered her up. And yet when I see another red-head – that's what I have to do."

"You loved her," I said.

He nodded.

"And now you carry a real neurosis."

The words flowed out, and I regretted them the instant they left my lips.

I edged closer.

"Dad, tell me: Why? Why did you leave her?"

He shook his head.

"I really can't say."

"Are you sure, Dad?"

"No," he responded. "It's too personal."

"It might help me."

"No. I can't say right now," he insisted.

I looked to the floor.

Why won't he tell me? I thought.

The thought of first meeting Nikki flashed in my head. He'd taken my brother and me to a darkened apartment – streaked mirrors on the walls. Then, as I sat playing with a toy on the couch, a tall, voluptuous nude woman emerged through a beaded divide.

I heard my brother call out behind me, but I sat frozen…

The dog's name was Rey, I continued thinking. Rey was Nikki's dog. She had been spayed – "Fixed."

Why had that thought – those words – come into my head?...

A stream flowed close to the campsite. I remembered the taste of the cool, fresh, running water in my mouth.

During the night we sat around a camp-fire. Other camping trips, my father had brought a short-wave radio and we'd listened to ghost stories. This time, he hadn't, so I attempted to invent one. At the end of my tale, Nikki threw her head back against the giant log that she'd been leaning against.

"Ugh," she sighed. "Why do children tell such stupid stories?"

Lifting her head again, she glared at me.

"Why do you think people have children?" she said.

I shrugged my eight-year-old shoulders.

"Maybe because they want company," I managed, smiling.

Looking upwards, she closed her eyes to the stars twinkling overhead.

"That's selfish!" she declared, angrily. "If you want company, get a dog!..."

The following day we took a long hike in the woods. Coming back I was smiling. All trip long I'd been looking forward to the promise of fishing, and my father had said we'd go just as soon we got back to camp. I could hardly contain my excitement.

Coming into view was a thin chain tethered between two trees on each side of the path and hanging about a foot from the ground. My brother ran ahead.

"Watch me, Dad," he called out. "I'm going to jump over it."

"Joe, you come back here!" my father shouted.

"Oh, let him be," Nikki retorted, annoyed.

Joe ran and leapt, but caught his toe on the chain. Though he'd tumbled in a heap, he quickly got up again, brushed himself off and continued his merry way.

"You see," Nikki declared, triumphant. "He fell. He's okay."

But I saw my father's features darkened, and knew what that boded, and felt afraid for her.

"Listen," he said. "When you have children of your own, you can let them run and jump and hurt themselves all you want. But these are my kids, and they'll do what I say."

"But as long as you've *'fixed'* yourself," he continued with a malignant hiss, "you have nothing to say about it…"

Arriving back at the campsite, I immediately set about assembling my fishing pole. Then, my father appeared and announced that we were breaking camp.

"Why, Dad?" I asked, anxiously.

"Because Nikki wants to go home!" he asserted...

During the drive, Rey sat drooling on me as a cold breeze poured through the open window. Then, around twilight, we stopped at what looked like a large log cabin. Inside, there was a large gathering of people: a man pleasantly conversing with another over a drink at the bar; a pleasant hostess led us between tables until seating us near a wooden stage; Nikki's features softening, as she pleasantly looked about.

"I used to dance at a place like this," she said. "I fell off a stage not very different from this one and hurt my leg."

Studying the stage, I imagined her falling, and felt sad.

"Is the reason you wanted to leave early because you were mad at my father for what he said to you?" I asked.

"That's part of it," she said.

I nodded and looked away, my eyes settling on a plain-looking woman at another table who sat smiling, pleasantly listening to the conversation and then breaking into laughter.

I turned back to Nikki again.

"What will happen now?" I asked.

"I don't know," she responded. "This might be our last supper..."

Not long after the trip, I overheard my father talking on the phone, saying he needed to find a home for Rey.

"What do you have to find a home for Rey for?" I asked.

"Because Nikki died!" he declared, bluntly. "She took a lot of pills and killed herself."

My eyes opened wide, I lost my breath, as my mind flashed to that last time with Nikki and what she said at the restaurant.

"Was the reason the two of you split up because of Joe and me?" I asked, timidly, looking for reassurance.

"That was part of it," he said, matter-of-factly...

Within the Cloister now, my father walked to the doorway, shoulders hunched.

"Why don't you admit that you're human," he sighed, his back to me, "and give yourself a break like the rest of us?..."

CHAPTER THIRTY-FIVE

"I need love."

Awakening, it wasn't so much desperation as a simple realization.

I'd loved a woman. No matter how I tried, I couldn't reconcile having left her.

Could it be that the abuses of the body go beyond the physical? Could it be they transcend something?

To rescind love? I thought. To rescind love...

At 5 AM I dialed the airport and made a reservation for a flight to Los Angeles that left at ten o'clock. Packing a couple books and change of clothes in a knapsack, it was as much as I dare carry.

If I reached the airport, I told myself, I'll notify work and family from there.

Then, the familiar sensation in my leg – the ripping pain, like a root dissecting its way through the tissues.

Would I make it? What would happen on the plane?

Beads of sweat formed on my forehead, as others coursed down my sides...

At the airport terminal my eyes wandered. A lone figure came into view – he was clutching at crutches in both hands, dragging his legs behind him.

My breath quickened, and a strange sensation crept up my spine; as I looked on with awe and camaraderie, and, all at once, the world became a blur...

CHAPTER THIRTY-SIX

Blue skies. Dry wind. A subdued greeting from mother – brother – the dog that beckoned me to play.

Dinner was prepared, and we sat at the table. My mother looked like a dejected child with wrinkles and graying hair, hunched shoulders and big sad eyes. My brother bragged, as he gulped down his meal, reclined in his chair, and proudly displayed his well-fed little belly. Continuing his soliloquy, he raised his voice, as though having to speak louder to convince himself…

The linens in the guest room hadn't been changed since my last visit. I'd foregone a visit during the winter holiday, thinking that there was no place for me there. Layers of dust clung to the furniture, and I was seized by a fit of sneezing. I went to my brother's room, and asked if I could sleep on the floor. Joe shifted his weight and opened a place for me on the bed...

During the night I was witness to every imaginable form of torture. Ashen figures with sunken eyes, gaping toothless mouths, and deep facial creases were set upon by demons, blessed with that vital glow that comes from a satisfaction in ones labors.

A victim lay prostrate on his back – a funnel forced into his throat, with yellow slime poured through it. His belly was tight and protuberant – as though at any moment it would explode – but his eyes were vacant.

My brother and I sitting on the couch, looking at the television.

"What warped, degenerate mind would dream up stuff like this?" I said.

I went to change the channel.

And woke up from the dream…

CHAPTER THIRTY-SEVEN

Ben arrived the next morning. I went to embrace him. He held himself rigid, and his eyes revealed a hopelessness that his smile tried to mask. Searching his eyes, I stopped myself. We piled into the truck like old times. The conversation was strained, at first, Ben talking about his job, the customers, the house, the economy. The pain surged in my leg. I asked him to pullover. He assured me that we'd be there soon. The pain was eroding my patience. No, I thought. Ben's too important. I can push through this.

Finally, he began to unravel.

"Michele and I haven't been intimate in months," he said. "I can't even remember the last time we'd made love when she was sober. Sometimes, I'll think about going out to a club some night, and, maybe, find somebody there." He lit a cigarette. "And, you know, I think that's just what she wants me to do. She'll go out with her friends to an all women bar... yeah, a strip joint. She says she's just going because her friends are... just to be with her friends, right? But, you know when she gets there, she shouting and hollering and getting into it just as much as the others.... I feel like she's deliberately trying to make me be unfaithful. I feel like the only way she can justify her feelings is for me to go out and do something to betray her. I'm not going to do it. I can't. I can't let my anger get the best of me. I can't go and do something I know is completely wrong."

"Have you considered, perhaps, a trial separation?"

"I can't," he said. "She wouldn't let me. She'd reject the idea as soon as I'd bring it up."

He hesitated.

"I'm sorry to be unloading all this on you. It's just that I don't have anybody else to talk to."

"No," I said. "I want you to tell me..."

The Greek-owned delicatessen was the busiest shop on the street. Casual, friendly, cordial – The owner directed cars in and out of the crowded parking lot. We smiled and relished the smell of fresh sourdough, and the sight of families sharing a meal. Our table was outside on the sidewalk, shaded by a veranda from the sun. Like the budding flower, forcing its way through the soil, and making its way towards the light, I had the feeling the life were being restored in me and breathed deeply and let out a satisfied sigh.

We rode out to Zuma and lay our towels on a patch of sand overlooking a couple of girls in bikinis on the beach. Walking towards the surf, we slid down a depression in the sand that brought us to the shore. I went running into the water, then dove head first into the waves. Looking up as I emerged, Ben was just behind.

Afterwards, we talked about our favorite books and writers, and then climbed the rocks, and looked out from the promontory.

"From a point like this you can understand why they thought that ships would go tumbling over some cliff at the end of the ocean," I said.

Ben stood beside me, nodding.

"Who would have ever thought the world was round?..."

CHAPTER THIRTY-EIGHT

Ben's father, Mr. Chan, had spent the day working outside the house, and looked weary when Ben and I arrived. But, seeing me, his eyes brightened, and suggested we go out for dinner. I jumped at the opportunity; but instantly regretted it, when I remembering how exhausted Mr. Chang looked.

Mrs. Chang's eyes lit up when she saw me at the doorway.

"Hi Mike," she said. "How have you been?..."

I gravitated to the bookshelves, pouring over the titles, and then picked up The New Testament.

"If there is a God, I say, 'I curse You!'... I curse Him!"

I'd been remembering my last visit with Mr. Vasquez.

"Are you interested in reading this one, Mike?" Mrs. Chang asked. "If you are, it's yours for the taking..."

We drove to Chinatown. Once inside the restaurant Mr. Chang underwent a dramatic change of character. He became refined and exuberant, ordering for everyone, insisting on fine Chinese liquor, and expounding on Chinese culture. Ben, the eldest son, enjoyed the honor of serving family and guest. Each course had to be consumed hot, Mr. Chang insisted, and devoured with gusto. Desserts were lavish.

"Mr. Chang, I just thought I'd tell you that when Gwen came back from China, she told me she enjoyed the trip with you very much," I said.

His eyes lit up.

"Yes," he said. "I think was a very gratifying experience – not just for her – was meaningful to me, too – for us both…"

CHAPTER THIRTY-NINE

In the evening Ben examined me.

"There are some people," he said, "who look like their work. You take one look at them, and you know what they've been doing all their lives. Look at a seamstress some time – They're usually hunched over because they're always bent sewing. Or a truck-driver and those big burly forearms from steering all day. They look like their work!"

He directed me to the mirror and pointed to the muscles of my upper back.

"They're massive!" he said. "They're literally massive!"

Then, he contorted his body and stood the way that I did.

"I can just see you hunched over your test-tubes," he said. "Carrying the weight of the world on your shoulders. I feel like I want to put you in a head-lock and straighten you up."

But wasn't this the way it was supposed to be? I thought.

"Posture makes such a difference in people," he said. "If you stand hunched over with your head to the ground, you're gonna be depressed. You're asking for it. I get like that sometimes – I know I'm doing it, but I do it anyway – because sometimes you just feel like that – You want to feel down sometimes."

"I'm lucky, though," he asserted. "Because in sales, you can't be that way. Before you go to a customer, you've got to take a deep breath, expand your chest, stand up straight, and give 'em a good impression of yourself. And the thing about it is, when you do that, you do feel better – You feel better about yourself."

I knew he was right. Still, wasn't there something noble about suffering? Wasn't there some humility in it? Hadn't this always been the way I looked at things? I mean, could I give that up without compromising my principles?

I want to live, I thought. God, forgive me. I want to live...

In the morning we gathered in the garden. Ben led me along a row of fruit trees. Large, exotic, colored oriental fish swam in the pond. Mr. Chang gave us some plums.

I perused the bookshelves a final time.

"Here's a book you should read," Ben said.

He handed The World's Greatest Salesman. Rifling through its pages, I found an inscription penned near the cover.

To Ben,
I know you will succeed in
everything you do.
Love,
Constance

In my mind I heard the voice that had echoed through the casino where Ben had been dancing. It had an easy quality – like a summer's breeze, or a lover's embrace. I turned to Ben.

"Who's that?" I asked.

"Constance."

"Your Constance?"

"Yeah," he replied. "I put her demo-tape on the loud speaker..."

They'd been together for several months. She had a noble quality about her – dispassionate and poised. Her love filled him – gave him strength.

The three of us went out after the show and found a club with music and dancing. The next morning we slept till dawn, then cooked omelets and drank more wine. In the afternoon we climbed Red Rock in the mountains just outside Las Vegas...

Constance had an abortion. She didn't tell Ben. They both had their careers, she said, and that had to be first in their lives now. He admitted that she was right, but would have liked to start a family – impractical as it might have been...

Ben betrayed Constance. He met a girl from the show at a club. They danced and then went back to her place. After, she told him that she loved him – in the same sweet voice that she told everybody else.

There were tears in his eyes when he came home to Constance. They cried all night, he'd said, scolding and then huddling together. She asked him if he loved this other woman? He said that he was in love.

"In that case," she said, "we can't go on."

She pulled herself up.

"We have to let go..."

"I could never take this book from you," I told him. "Don't ever let it go…"

CHAPTER FORTY

Traveling the grapevine, the surrounding mountains seemed a living tapestry. Rugged and full, with crevices that appeared breath, and lined by grass that resembled fur from some mammoth sized creature.

Then, the San Joaquin Valley came into view, revealing a panorama of vast farmlands, and earth so rich that the soil moved out to permeate the air, spreading into the horizon in all directions. Golden grass lay on level ground that the herds of cattle munched on, and then miles of tilled land that stretched along the highway.

As the bus moved along the highway, the migrant farm worker sitting beside me commented that I seemed to be shifting about and asked if I were uncomfortable? I told him that I'd hurt my leg and the long ride was difficult for me. He looked out into the vast fields, his eyes vacant.

"That is the life," he said…

Bubbles' father, Mr. Vasquez, was waiting at the terminal. He jumped up in front of me, enjoying my startled response to the absence of his familiar mustache.

"My wife wanted me to shave it," he said. "She said the gray in it was making me look older."

Walking to his truck I thought about what he looked like at the time of our first meeting. His daughter and I had been sharing an apartment then. It was summer, and I'd had my shirt off and the fan going; listening to music in my room, and strumming my tennis racket like a guitar. Then, a knock came from the door. Not thinking, I called out, "Who is it?"

Then, turning to the doorway stood the two largest Mexican people I ever saw.

"Mike, I want you to meet my parents."

The voice was Bubbles', but I didn't know where it came from, as my eyes were fixed on the couple in front of me.

The couple was staring at me, too. Both were nodding, quizzically. Mrs. Vasquez's eyes narrowed, and from Mr. Vasquez I heard a subdued rumble.

"Ay," he said, taking in my 'guitar' and semi-nudity. "Yeah."

Creeping forward, I felt like a mouse sniffing at a piece of cheese – excited, yet cautious.

"You need to lock the door," Mr. Vasquez advised. "You just get a key made, and you can go in and out whenever you want to, and not worry about thieves…"

But I never did get that key made, and the door remained unlocked and open all summer. We were wild and free, and no one and nothing was going to change that...

On the way to the house I asked Mr. Vasquez if we could stop at the florist?

"I'd like to buy your wife some flowers," I said.

He stopped.

"Don't spend more than a dollar," he said…

He asked why I hadn't driven? I told him about the problems with my leg.

"What you really need to do is see a chiropractor !" he said. "You see a chiropractor – forty dollars – he fix you – fix you right up – fix you real good. He run a finger down your back – 'Right there!' he tell you – don't need an X-ray or nothing. He just fix you. Fix you right up… "

At the house Little Davy was anxiously waiting.

"Hi Mike!" He hugged me with his taunt, wiry arms. "This one's for me... This one's for Mom... This one's for Tom... And three more for our new foster kids..."

By the time he'd released me, I had to catch my breath.

Little Davy was bristling with excitement and had the rest of our day planned. First, a movie. Then, video games. Then, dinner out. I would have liked to wait for Bubbles' younger sister, Nellie, but didn't want to dampen his enthusiasm, and the two of us left at once.

In the growing shadows Davy and I sat outside the mall. His responses to my inquiries were usually limited to single syllables – "Good", "Fine", "Okay." In between, he frequently broke into uncontrollable fits of laughter.

He had never met his real father, who had since died. With each time his mother, Melissa, had re-marry, he'd acquired a different name – Davy Vasquez. Davy Hannover. Davy Hoosier. Davy X.

He'd been four years old at the time of my first visit. On our first outing he and I had accompanied his aunts to the mall. At a clothing store Nellie and Bubbles looked through the dress racks; I followed little Davy around the store.

He made his happy way between the aisles, entertained by anything that came his way. Then, he tried to put his legs through a clothes hanger, and went crashing to the floor. Crying more for the unfair trick played on him than injury, I lifted him into my arms. The girls made their way over, and, tickling him, they called him, "Little faker." He responded with bursts laughter...

Mrs. Vasquez and Nellie picked us up in front of the sidewalk. Nellie looked tired, and unlike her usual spirited self.

Mrs. Vasquez prepared a place for me in the guest room, as she'd done for years.

Peering into Nellie's room through the crack in the doorway, she and Davy sat on the bed watching television. She offered me a place on the bed. I told her that I was comfortable on the floor.

After a few minutes Nellie lay out on the bed, adjusted the alarm clock, and gave a sigh goodnight. She lay smothered in the bed. Still, as Davy and I finished watching the comedy, her laughter still joined our own. I thought of how her boyish expression that changed when she laughed and became aglow. I looked over, then lay my head back and smiled...

CHAPTER FORTY-ONE

At dawn Mr. Vasquez had already risen and was preparing waffles and coffee.

We held forth, convening on all matter of state; engaged in discourse that ranged from crime and the recent riots, to disease and epidemics, and, finally, power, privilege and persecution in the world at large. Little Davy had long since come and gone, taking a waffle and sleepily consuming it, before leaving the room.

Nellie appeared tired and expressionless. There was an unnatural pallor about her lips and face, and her hair hung limp upon her shoulders. She looked smaller – vulnerable. She held herself out to be hugged, and then left for work. In the kitchen I noticed that she'd forgotten her lunch. I ran to catch her and her expression livened...

I walked about the park near the house and looked out at the peaceful blue skies, humble residential homes, open fields, and barren brown landscape. Mr. Vasquez led me through his garden. Their many dogs leaped on me, the puppies gnawing at my ankles. Mr. Vasquez dusted the mud off his back, picked some strawberries, and offered them to me.

As we sat on the porch Mr. Vasquez pointed to the different birds. In particular was the mimic – a brilliantly colored bird with red-faced with blue-outlines, that flew with the precision of a humming bird, and the grace and swiftness of a sparrow.

"It lays its eggs in the nests of other birds," he said. "When the eggs hatch, the other birds take care of its young…"

A casually dressed man carrying a duffle bag appeared at the gate. He was accompanied by two small children.

"Hello, neighbor," the man said. "I just wanted to introduce myself. I'll stay on the other side of the gate."

He had a glazed look, and never made direct eye contact. The small talk lasted for several minutes before finally announcing that he'd for the local chapter of Jehovah Witnesses, and was making his neighborhood rounds. He asked Mr. Vasquez if he'd read the Bible, and then launched into a lecture. The man's two children stood timidly looking up at me in intervals with shy, ingratiating smiles.

"For me," Mr. Vasquez interrupted, in his husky, domineering voice, "doesn't matter what religion person is – what they believe. They good neighbor – is enough for me. That all that matter. Good neighbor – doesn't matter what else you believe – is good. You not steal from me – you not break into my house – you not harm my family – you alright. What you believe in – doesn't matter to me.

"I have a brother. He come to my house – he bring his Bible with him – he says he has to carry it with him – everywhere he goes. He used to smoke – drink a lot. Now, he says he's found God – cut it all out! I say 'Good! You don't drink. You don't smoke. Good. I like that. Is something good.' But when he bring that Bible into my house – and he start trying to lecture me – and converting me – I say I don't need that – not in my house. You show me respect in my house..."

A year before we'd been spending the evening in the open air as we sat perched atop the treehouse. As well as me, Mr. and Mrs. Vasquez were entertaining his older brother, Juan. Juan was a born-again convert, and held strong opinions on the matter of faith and religious doctrine. There was the usual banter, but, suddenly, Mr. Vasquez's face twisted in rage.

"If there is a God, I say, 'I curse You! I curse You!'"

"No, Chavo," Mrs. Vasquez interceded, trying to calm him.

But her attempts were to no avail, as Mr. Vasquez's expression only darkened.

"I curse Him!" he continued. "That my daughter, Sylvia, can't have children – That she had to have that operation. I say to Hell with Him! I curse Him!..."

The Jehovah witness, appearing anxious to continue on his rounds, hurriedly voiced his agreement that the brother's devotion was probably 'going too far', and said his goodbyes, then sped off with his children.

Opening the New Testament that I'd acquired at the Chang's, I read the passage in front of me.

"Obey the commandment – Love thy neighbor – and everything else will follow..."

CHAPTER FORTY-TWO

A car pulled into the driveway: Melissa had arrived with the three foster children. The kids spilled out of the car and came rambling into the house. All three had broad builds with droopy features (like endearing bloodhounds), ages one, two and three.

The two-year old was outgoing and made a B-line for anything that wasn't nailed down. The one-year old was thoughtful, scrutinizing everything and everyone in his way. A peculiar pathos seemed to trail the three-year old, his downcast eyes appeared terror-filled, and he made a deliberate effort to not stray far, as though expecting a beating at every turn.

Later, Davy conveyed that the children had recently been taken away from an abusive couple and described their condition at the time that they'd been received.

"Their eyes were swollen and bloody. There were bruises and rashes all over their skin. Pot marks. Burns from cigarette butts. They were swimming in filth…"

Mrs. Vasquez and Melissa returned carrying a bag full of photos from the wedding of the Vasquez's eldest daughter, Sylvia. They sat around me on the couch, going through the pictures one-by-one. There was dancing and mariachi players, and, in the center, a joy-filled Sylvia, looking triumphant.

Then, Mr. Vasquez came bounding back into the room.

"These are the pictures I like!" he asserted.

He pushed a package of photos into my hand. At first, I saw what appeared to be a lot of lightly clad women on a stage.

Then, looking closer, I saw Nellie, and realized that these were pictures Mr. Vasquez had taken at a recent beauty pageant. Nellie had worn the same two-piece bathing suit as that last time we'd taken Davy to the waterslides. She looked smart, spirited and natural.

"Oh, Chavo!" Mrs. Vasquez called out, smiling.

"I like these pictures!" Mr. Vasquez insisted, good-naturedly. "I like them better than the others!..."

Little Davy herded his foster brothers, and forced them to sit together in a crowded armchair. The children were uncomfortable, and three-year old began to sob.

"Davy, will you stop!" Melissa cried in a strained voice.

Little Davy said he was only doing what he thought was right – he was just trying to help. Melissa stood towering over him; Davy's eyes filled with tears as he barked back. Finally, he flew out of the house, slamming the door behind him.

I excused myself and followed. He sat crumpled up on a chair in the porch. Still sobbing, his face displayed utter hopelessness and frustration.

"It's tough being a big brother," I said. "I always hated it."

Davy grimaced. The two of us remained quiet on the porch, till the tears in his eyes dried; then, he opened the door, as though to signal the resolution of the conflict and his willingness to move on.

The rest of the family came out. Melissa sat beside me on the steps. Mrs. Vasquez kneeled on the grass; adorn in a white dress, with a radiance that eclipsed all the flowers that surrounded her. The three foster children scurried about the yard; making a bee-line here, then another there; the two-year old leading the way, his brothers rapturously in tow, as though they'd found a paradise in the small fenced-in enclosure. Mr. Vasquez sat on his chair on the porch. A cockroach crawled out of the yard, and made its way up his leg.

"That's disgusting!" Melissa cried.

Mr. Vasquez nonchalantly slapped it to the ground, then gingerly crunched it under his boot – checking underneath with each crack.

"Daddy?! Kill it! Kill it, Daddy!..."

"I don't want to get it all over the porch," he said.

Taking a handkerchief from his pocket, Mr. Vasquez scooped the insect's remains, then chased Melissa around the yard.

"No, Daddy!" the enraptured cries.

Mrs. Vasquez shook her head, and vainly called her "Chavo" off.

Little Davy shook his head, and his look reminded of a book that started, "When I was born, I had two children... my mother and my father...."

Love and hate meeting as one. Running from the arms you long to hold you. Breaking away from those you desire most. To probe and penetrate the unfathomable darkness that is the human soul, from which there is no return, but only descent into an abyss which offers no guarantee of light, and for whose whole purpose might be to quench the light. But to go on probing the darkness, nonetheless, by chance to come upon something tangible – something of substance – of heart – and to be enraptured with delight, till flowing into the current that embodies the human affairs till caught up and swept into the waters, in hopes of not being dashed against the rocks, and, like the stone that's smoothed by the passing stream, so to know another day, and probe again those last refuges of the lonely human heart to which we cling by loose threads that bind and hold and meet in love and make us whole...

CHAPTER FORTY-THREE

The foster kids were packed back into the car. Melissa stood with her back towards me; eyes looking into the distance; shoulders soft and slightly hunched. The heat emanate from her body was familiar. She turned to me, and pulled me close. There were tears in her eyes. We kissed. I watched her pull away...

Mrs. Vasquez embraced me.

"Thank you for not forgetting us," she said.

Mr. Vasquez drove me back to the bus station. He pointed to a factory along the way.

"Twenty years I worked at that place," he said.

Looking over, I thought of his children.

Then, his features softened, as though he were deep in thought.

"Not long ago I went over to Melissa's," he said. "She and her husband were having a fight – bad fight. He pulls out a sawed-off shotgun – Holds it to her mouth. Says, 'Shut up. Shut up. Or I'll kill ya.'"

He looked away.

"When I was young," he continued, "I used to wear chains on my belt – carried a knife – never backed down from a fight. Now, I see my daughter like that – this guy pointing a gun at her – gonna blow her head off. All I could do is sit there and watch..."

Ascending the stairs to the bus depot, I looked back, as he waved to me from the truck.

"Come back and see us," he said...

CHAPTER FORTY-FOUR

My mother picked me up at the bus depot. We went for pizza, and, at the restaurant, she stared out with deeply sunken, darkly ringed eyes.

"How is your pizza, Mom?"

"Good."

"What did you do yesterday?"

"Nothing."

"How is Joyce?"

"Fine."

Years before she'd had a nervous breakdown. I'd been in Maine. My uncles told me that they'd take care of it. Returning home, though, she was only the shadow of the person she once was.

But she'd been unusually capable during this visit? She'd made me dinner, picked me up. How was it she'd been able to do so much?

Packing my things I came across a photo on her dresser. A picture of me that she'd taken it a year ago. It had been blown up, and framed. I looked handsome, spirited, energetic. What did she see?

Then, a surge of emotion gripped me, and I breathed like a man emerging from underwater to the surface after having been submerged for an interminable time...

"I love you very much," she told me at the airport. "Tell me whenever you want to come home – even if it's just for the weekend. You're always welcome...."

The airport was dark, and waiting in the terminal I thought about my visit. I'd long run out of medicine, but the pain didn't seem to matter; and I couldn't help the feeling the future looked bright, the wounds would heal, and the door that closed would open another.

As long as one moment of life existed, I thought, I was going to find a way to live it…

Part Four

The road to healing took me to Dr. Amin, Bioenergy, Chi Gong and Elizabeth.

Then, Bubbles made a surprise appearance…

CHAPTER FORTY-FIVE

August, 1995

"When I hold you," Liz began, "I experience a feeling of joy and friendship – like we were children, playing on a sunny day in an open field. It's an almost mystical feeling. When I meditate, I see an image of you – you're always naked, sitting in a lotus position; there are swirls of energy about you, and you either have your eyes closed, or you're smiling at me with your brightest smile. It's this image that my eyes call to mind before I can smile into the rest of my body. And in the times that I'll be upset and won't be able to find my way, I'll feel like there's something that I've forgotten, and I'll call up your image and be able to move forward again."

She looked at me.

"Michael, do you still think I have a funny smile?" she asked.

I narrowed my gaze.

"Why do you ask that?" I responded.

"Because that first time I met you in the clinic, you said that my upper lip appeared different somehow. You asked Dr. Weber about it, and he said it was because of the faulty way that I used my jaw in order to accommodate a misalignment in my bite."

Tears formed in my eyes.

"No," I said. "I don't think you smile funny anymore."

"Why are you crying?" she said.

"Because when we first met," I said, "it seemed like nothing came natural for you, and everything was forced – even your smile. And, now, it isn't that way anymore."

I find you truly beautiful, I thought without saying.

"Everything had been an effort before," she said. "When you're sick, everything is."

She looked at me.

"Michael, do you believe that I love you?" she asked.

I shrugged.

"I know you love me," I said.

"I remember the day I went to your home," she said. "I had just parked my car at the Cloister. When I got out, I saw you walking towards me. I turned away, and then you ran to me. You pulled me around and held me close to you. There had always been a pain in my chest. I'd always had it. It was just a little pain, but it was always there. But the instant my chest made contact with yours, it went away."

She took a paper from her bureau.

"I had this thought about you while I was doing Chi Gong this morning," she said. "Remember we were talking about your heart pain and I thought you might try an opposite approach from what would seem the intellectually logical one? That is, when you have heart pain, instead of recoiling from it, or delving deeper into it to discover or face the source of the pain, or even instead of nursing your crushed heart, what if – sometimes – you tried using it to love, exercising it, teaching it to love, trying out different feelings or aspects of love, just one at a time, slowly, little by little."

In this my heart so full of you,
Is a tiny place where ever-present night dwells without stars.
If you will light me a candle and hold my hand,
I will go and see what is there.
Then, maybe, I will feel the warm glow of the bright colors
That dwell,
In the expansive chambers
Where ever-present day enjoys the delights of life

With you in my heart…

"You've said that you tend to get the pains when you're getting love, affection or positive feelings from someone else," she continued. "Perhaps particular loves that you didn't get or were denied as a child and the sudden receiving makes you remember the pain of not getting before. So, trying to exercise your own ability to love at that point might shift the emotional emphasis from a heart that sees itself as a destabilized victim to one that sees itself as an independent functioning lover of itself, yourself and others. The shift is from a weak position to a strong one..."

At the Vasquez home, Bubbles' father, Mr. Vasquez, was driving his truck into the garage. Elizabeth was with me, and I presented her to him.

"Ahh," he responded, dismissively. "You don't want that."

I turned, worried for Elizabeth.

Mr. Vasquez brought out a detailed pencil sketch of a stalk of corn and asked what I thought of it? I responded that I was impressed with the workmanship and detail in the drawing. But he was not content, saying he meant to give it to me as a present, and Hanukah was tomorrow, but it wasn't ready yet.

"Then, I missed Thanksgiving and Halloween," I lamented.

"No," he responded. "That's not true…"

Awakening from the dream I felt a pain in my chest. It wasn't the usual piercing heartache of missing Bubbles, but, rather, a kind of numbness.

What I wouldn't give to see her again? I thought...

Volunteering with Dr. Amin at his Alternative Medicine clinic I accompanied him as he evaluated patients. As a patient myself, he'd helped me with my leg pain, and from the start I'd been fascinated with his use of a healing technique called bioenergy, in which he was able to scan the body for energy imbalances and then correct them. Now, as a medical student and scientist, I wanted to learn this technique...

Near the end of the day, I felt tired. Then, Dr. Amin led me into the room of a woman with bright smiling eyes and shy way about her, and all the exhaustion I'd been feeling fell away.

Chatting with Dr. Amin, the patient had a chipper, curt way of talking, with a tendency to gush and laugh as she spoke, even at dark humor. It all sang of a long forgotten melody – All those beautiful sounds that I missed and had always for me held such an attraction – It was the sounds of Bubbles.

"I have numbness between my fingers," she said of her condition. "They ache."

"Gina is a flutist," Dr. Amin added, "so problems like these pose an occupational hazard."

He examined the patient's back.

"She has a rotation between her second and third lumbar vertebrae," he said. "Feel it."

Palpating her spine, the vertebrae felt prominent and out of alignment.

"Now, see what you feel when you scan her for energy disturbances," he instructed.

Moving my hand over her back, I perceived the sensation of an energetic breeze blowing from between her shoulders. As in the first time I performed bioenergy (and every time since), I was immediately imbued with a feeling of awe and reverence, so to slow me down, center me, take away all peripheral thoughts and concerns, and make me want to explore the person's condition more fully. Before learning this technique I'd only had this feeling once before – That day with Bubbles in the fields beyond the Physics Building. Now, with bioenergy, I was experiencing it with each and every patient.

"The energy is interesting," I commented. "Over the right shoulder it's as though the energy is directing my hand inwards, whereas for the left, it's pushing my hand back."

"The fascial pulls are tugging her into the position of her vocation," Dr. Amin explained. "Holding a flute."

"Gina overdid it last summer," he continued. "She took too many jobs at outdoor events. Working days and evenings, she overstressed the muscles."

Just then, the muscles in her shoulder spontaneously jumped, so to signal some significant myofascial release, and when I looked for the energetic strain again, I couldn't find it.

"Look at her back," Dr. Amin said.

Examining her vertebrae, they no longer felt out of alignment.

"It's as though everything just fell into place," I said, confounded. "How did that happen?"

"A person who's been injured taps into your body to get relief," he said.

"What?!" I responded. "How does my body know how to heal hers?"

"I don't know that it does know," he said. "The body just does it..."

"There's another way to perceive what's going on in a patient," Dr. Amin told me.

We were in an exam room with his patient, Mr. Penn. Dr. Amin instructed the patient to lie with his back against the examining table.

"Mike, stand at the foot of the table and grip the patient's ankles," Dr. Amin said. "Now, try to open yourself to any new sensations."

All at once I experienced a fatigue and heaviness along my chest and abdomen. Then, a pain in my lower back developed. It was sharp pain, mostly on the right. I tried to get past it, but it seemed that I was stuck there, and my muscles were clamping down. I looked at Dr. Amin.

"What is this?" I asked.

"Interlink," Dr. Amin said. "It's a means of energy transfer... The body is crying out for a connection. It wants to tell you what it needs."

I clamped my eyes shut and tried to tolerate the feeling.

"Mr. Penn, where is the majority of your pain?" I asked.

"In my back," he responded.

"Which side?"

"On the right."

I hung on, but the discomfort was intense.

"Dr. Amin, I can't seem to get beyond the feeling that I'm getting from his back," I said. "I seem to be stuck - I can't move any lower. Is there anything that I can do?"

"Let go of his ankles," he responded.

I took my hands away; within an instant, the intense tightness encircling my lower back was gone. I looked about, startled.

"That's it?" I asked. "It's that simple?"

"That's it," Amin responded, casually. "The body's said what it needed to tell you."

"No?!" I thought.

I clasped the patient's ankles again - the fatigue and heaviness in my chest reappeared, then the clamping sensation in my back. I let go - it evaporated. I looked to Dr. Amin.

"It can't be?..."

In the evening Elizabeth invited me to meet her mother, Lena. My first impression was that she walked with difficulty, wobbling from side to side.

"Mom, how long have you walked this way?" Liz asked.

"I couldn't say," Lena responded. "For years... I don't have any pain. It's just difficult for me to walk straight."

"Mike, what do you think could have set it off?" Liz inquired.

I shook my head.

"Could be a lot of things," I responded. "Often when Dr. Amin works with patients who have walking difficulties, he asks about surgical scars and pelvic trauma, saying that's often the source of their difficulties."

Lena returned to the couch and sat, reflecting.

"Maybe it was from having six children?" she commented. "I also needed an appendectomy and cholecystectomy."

"Mike, why don't you try doing bioenergy on her?" Liz urged.

Scanning Lena, the energy radiating from her expanded outwards, so to seemingly fill the apartment.

"I've never experienced anything like this before," I said.

Following the energy's path took me across the room, into walls, through windows, and, at times, required that I get on a chair to reach high enough.

In the meantime Lena had fallen asleep.

"She has narcolepsy," Liz said. "Should I wake her?"

"No, let her sleep," I insisted. "I feel embarrassed, as it is."

Lena awoke shortly after I'd finished.

"I'm sorry that I fell asleep," she said.

Then, she got up, and as she walked across the room, Liz jumped back.

"Mom, you're walking straight!" Liz called out.

"I am?" Lena asked.

Walking back to the couch, there wasn't a trace of her former wobble, and, unlike before, she moved as though it were no effort at all.

Liz turned to me.

"Mike, that's amazing," Liz said...

Elizabeth had an appointment with Dr. Amin the following day. I accompanied her to the clinic, and Lena came, as well.

"Hey, guys," Dr. Amin said, cheerfully entering the exam room and greeting the three of us. "How are you doing?"

"Not so good," Liz responded, looking out of sorts. "The two sides of the brain didn't feel even."

'Don't feel even?' I thought. That's an odd thing to say.

Dr. Amin, however, was unfazed and went about examining Elizabeth's reflexes.

"Her reflexes are all unequal," he asserted. "The right biceps is different from the left. The right knee is different. The right ankle is different from left ankle. It's that way all over the place."

Taking the reflex hammer, I repeated his examinations.

"Yes, it's like it's dead on the right side, and hyper on the left," I said, nodding. "What could account for that?"

"Did you ever have a head injury?" Amin asked.

"She had a concussion when she was an infant," Lena inserted. "She was swinging on a rocking horse and fell off. She landed so hard that she lost consciousness. When I took her to the emergency room, the doctors insisted that she stay overnight for observation and testing. The next morning, though, they said that she was alright, and I took her home. She didn't appear then to have any problems, but I was always concerned there seemed to be something different."

"Well, I think that she had some kind of a central trauma," Amin responded, "because there's something going on way up in the brain. Let's check your cranial motion."

He cupped her head in his hands.

"Part of the temporal bone on the right side isn't moving well," he said. "What happened on this side?"

"That's where she fell," Lena said.

Dr. Amin looked at me.

"So, confirmation," he asserted. "Look, I think the fall stunned the temporal joint. Next, at this point, we have to straighten out the temporal bone in the head, because, essentially, that's what's going on."

"How do you do that?" I asked.

"Cranial sacral therapy," he responded. "If you were to take a look at the bones of the skull, you would see that they have an articulating surface and are not fused. So, if you look at the form, you have to conclude that they move, and have the ability to permit cranial motion, and it's a pumping mechanism. The cranial rhythm is the breathing cycle for the nervous system. It originates at the skull and traverses the entire body. Where there's an uneven expansion of the cranium, it means there's an unresolved trauma."

Closing his eyes, he focused.

Lena edged closer.

"Dr. Amin, I work at the school as a child psychologist," she said. "Have you ever treated children with learning disabilities?"

"Yeah, sure," he said. "I have one patient whose son is autistic. Mike, you remember Mr. Helfman?... Well, this is his autistic son. I gave the parents a few exercises to do with the child to improve the kid's cranial motion, and, after that, everything changed. His whole affect changed. Now, the son's improved so that he's got to a point where he's able to function with normal children now. This was an autistic kid. The same child who was incapable of tying his own shoe laces earlier.

He's started to talk clearly, and wasn't staring out into space anymore.

"And he's doing very well. He's living like a balanced child. Mr. Helfman says that he's expanded his interests. He never had an interest in anything. He never wanted to try anything. Now, he trying new things. He's doing new things. He's running around. He's asking to learn new things. His mother home schools him. She says, 'He's asking me to teach him, and I'm having to keep up with him now, instead of the other way around.' And this was an autistic kid.

"In fact, they put him in summer camp. This is a kid that before that barely able to speak. Now, he's speaking, and getting along with everybody.

"This is what's interesting, though. At the beginning of the camp, they had a party for everybody. They're all sitting at a table, and one of the girls starts to choke. She starts to point at her throat, and everybody is sort of like staring at her, as she's turning blue. And this kid, with autism, walked over behind her, grabbed her under the rib cage, did a Heimlich maneuver. The pizza popped out across the table, and he went back to his chair, sat down, and continued eating his pizza. Everybody there was just shocked. The kid got a gold star for it."

Lena nodded.

"There is so much potential out there," she said. "Just think what could happen if other autistic children got the same treatment?"

Dr. Amin stood.

"Alright, we're done," he asserted. "Liz, you rest here on the examining table for a few minutes. Lena, stay with her and let me know if anything happens, alright?... Mike, come with me, I have an interesting case I want to show you in the other room..."

Exiting the other exam room Dr. Amin and I were still talking about the patient we'd just visited when I noticed Lena standing in the hall.

"I think that you ought to see something," she said.

We followed her back to Liz's room. Inside, Liz looked out, as though captivated and confused at the same time.

"What's going on, Liz?" I asked.

"I...," she began.

Then, before she could produce a word more, her eyes spontaneously began darting back and forth in her head – the movement so fast as to be impossible to be voluntary! It was as though the muscles behind her eyes were in a violent tug of war.

I stepped back and turned to Dr. Amin.

"What's happening?" I asked.

"The brain is reacquainting itself with the muscles of the body," Dr. Amin responded. "The head injury that she had when she was young occurred at a time when her brain was still developing. The trauma shut down the development of more sophisticated pathways. It had to resort to the less sophisticated pathways that were already established. But these were pathways that should have been passed a long time ago, like early milestones in development. She's been relying on these old pathways all this time. Now, the sophisticated portions of the brain are reacquainting themselves with the body."

Elizabeth's eyes stopped moving. She breathed a sigh of relief.

"Whew," she said.

But, then, her body shuddered, and her jaws began moving, as she bit up and down uncontrollably, with the same unworldly quickness.

"Jesus!" I exclaimed. "How long can this go on?"

"Who knows?" Dr. Amin said. "It can take the body a while to adjust."

"How long is that?" I asked. "Hours?"

"Sometimes, it takes days," he said.

"'Days'?!" I exclaimed. "Can the body take it?"

"It knows to stop and rest when the person gets tired," he said.

Her jaws began biting in all directions.

"She's going to crack her teeth!" I declared.

"No," Dr. Amin responded. "The body knows to adjust for things like that. It isn't going to do things to injure itself."

The 'reacquainting process' continued, progressing from the muscles of the head downward.

The pattern was always the same. It began a shock wave ripped through her body, and heralded the movement of the reacquainting process in a new muscle group. Then, the movements began.

The process followed a seemingly orderly progression, sequentially moving from the muscles of her neck, to her shoulders, back, arms, and legs. The movements were too fast to be voluntary, and seemingly too precise to be involuntary.

It was apparently as Dr. Amin had described - the brain was reacquainting itself with each individual muscle fiber; familiarizing itself with the muscle's every action and movement for the first time; learning the correct circuits to voluntarily operate the muscles all over again.

At times, the movements seemed violent, as the muscles stretched the limbs to the extent of their capability. It also seemed like certain muscle groups were receiving more attention than others, particularly those involved in movement of the eyes, jaws, neck, and thumbs. At other times, Liz's body seemed to twist and assume odd positions, as it readied itself for another reacquainting session.

"Here we go again," she said.

Her left shoulder began contorting. The rapidity and containment of the gyrations suggested something beyond voluntary control. Then, her left foot began contorting - first, at the ankle, then at the great toe. Ultimately, it moved to the hip. It had the effect of looking like Liz were a marionette, being pulled back and forth by strings.

I turned to Dr. Amin.

"Were you focusing on certain muscles when you worked on her?" I asked.

"No," he said. "I just found the restriction in her temporal bone and relieved it. After that, her body's been in control."

"And correcting that one little bone is responsible for all this?"

"It's not a minor maneuver," he said. "If my car doesn't start, it's because there's a bad connection between the ignition and the battery. Someone who knows what to do can just come in and knock on the turbo, and, the next thing you know, the car can start... Well, if you don't understand what's going on, then it's a miraculous event. In other words, this whole system just got going because this man made a little tap, like that. It's the same thing in medicine. Her whole brain, her whole body, her whole nervous system - Everything just got kicked into place because that 'one little bone' was impeding the good function of the brain. Now, suddenly, it's no longer an issue and the brain can function.

"So, it looks like miraculous things are happening, but it's really nothing more than basic physiology, if you apply these principles. You see what I'm saying?"

Between contortions, her body stayed still for a long time. I didn't know whether it was because the subconscious was integrating what it had just experienced, or if it was scanning the body for the next distinct set of muscles that it wanted to initiate into the reacquainting process. The body moved her like an automaton. It was like her subconscious was connecting to a new set of circuits and looking for the best way to master the gears...

The 'reacquainting process' finally over, I assisted Liz off the table.

"It's like getting use to a complete re-wiring job!" she exclaimed. "With each step, though, I'm more familiar with how the limb works."

Dr. Amin checked her reflexes again.

"All her reflexes are bilaterally symmetrical and normal," he said.

"No," I thought. "It can't be."

Taking the reflex hammer, I repeated the examination; her reflexes were equal!

"I feel like I'm being held up by puffs of air," Liz continued. "At my back, my front, by my sides. I feel like my mind hasn't connected with my muscles yet…"

In the evening Elizabeth stayed home recuperating, so I went to Chi Gong practice at the local high school alone.

Inside the gymnasium Master Shang's assistant, Don Ko, led the group in its preparatory bows.

"Begin," he signaled.

As the practice commenced, energy flowed into my hands.

Then, something happened. My body was physically moved to walk in a certain direction. It felt like some force had taken hold of me, and my movement was outside of my conscious control. The movement was light, and completely unlabored - as though walking on air. It was like nothing that I had ever experienced. This time I was not following the energy, but, instead, the energy directing me...

"It was just a few short steps," I told my friend, Al Lee, at the conclusion of practice, "but it felt like the most extraordinary journey of my life."

"I'm happy for you, Mike," he replied. "That's what Chi Gong is supposed to feel like. Some practice for a life time and never experience that."

Don pulled me aside.

"Master Shang has chosen you to become an assistant," he said. "Though you have not been practicing Chi Gong for very long, he thinks in your heart are all the qualifications to be an assistant."

He smiled, broadly.

"What do you see as your mission in life?" he asked.

"I want to continue my bioenergy training with Dr. Amin," I said, "and cultivate my ability to serve as a channel for healing - a conduit for correcting energy imbalances."

"Through Chi Gong," he responded, "you will become an energy conductor... A deliverer of that energy."

"I don't know that I want that," I said. "It sounds forceful. I just want to help others reconcile their energy blockages. That's why I entered Chi Gong... So that I could rid myself of the blockages that were getting in the way of being a better healer."

"How successful have you been with curing patients using bioenergy?" he queried.

"I've had many successes," I said, "but, also, failures."

"What failures?" Don asked.

"Elizabeth, for example," I responded, timidly.

"Why do you think that you haven't been able to help her?" he inquired.

"I think it's because I can't let go of something," I confided.

Breaking off, I sighed, lowered my head and rubbed my eyes.

He nodded.

"Emotion," he said. "Emotions get in the way..."

Leaving the high school I decided to spend the night with Liz, feeling that if she had this reacquainting process happen to her again, I wanted to be there to help her.

Arriving at her apartment, she smiled.

"That thing happened to me again," she said. "But this time it was less intense and different from what happened at Dr. Amin's office… I was sitting in a lotus position, practicing Chi Gong, and my hand started knocking on my chest, and I was spontaneously saying, 'Hate, anger, frustration – Leave my body. Hate, anger, frustration – Out, out, out.'

"Then, a new chorus started: 'Love, compassion – Don't hate. Love.' And it felt as though the hatred and frustration was leaving me through my feet."

Nodding, I thought about experiences at Dr. Amin's clinic, in which after treating a person energetically, I perceived a flow of energy to the ground.

"And I had an interesting insight, too," she continued. "The insight told me that I had so many organs that were blocked and malfunctioning that the free flow of energy in my body was restricted, and that why I feel hate and frustration more often. It's like the chicken and the egg story… Which comes first? Do your organs become blocked because you feels so much hate and frustration? Or is it that because you don't feel good as a result of faulty health and malfunctioning organs that you become more commonly flustered and frustrated?"

Liz rubbed her sides.

"Miker, would you do some bioenergy on my back?" she asked.

Searching for energy imbalances, I moved my hand behind her. Perceiving what amounted to the feeling of an energetic breeze blowing at my hand, I followed the energy till the feeling dissipated.

"I think that I can feel where your hand is," she commented.

"Where is it?" I asked.

"It's at my shoulder," she said.

"Which shoulder?"

"The right," she said, correctly.

I moved it somewhere else.

"Now it's at my sacrum," she said.

"Where is it now?" I asked.

"At my head," she said. "Is it there?... I can feel the energy coming from it. That's amazing!"

Tears spontaneously welled up in my eyes. She turned.

"Why are you crying?" she asked.

"Because I cherish you so much," I said. "For the past weeks I haven't known exactly what to do. It doesn't seem like you're getting better, and you're losing weight again. I don't know what to do."

"Mike, this is the happiest that I've been in years."

But, as I scanned her again, a spreading feeling of tightness expanded across my chest. There was something getting in the way of my being able to help her, and I couldn't be sure what it was?...

Returning to my apartment, the light on my answering machine was blinking red with over ten messages from the past days I'd been at Elizabeth's.

"Mike, it's Bubbles," the last one began. "Hey, I'm coming to Philadelphia for a continuing education seminar. I heard that DC is near there. Have you seen Philadelphia yet? Maybe you can come by and we can explore the city? I'm on a tight schedule, but if we meet on Tuesday, we can spend the day. Let me know soon."

'Tuesday'? I thought. That's tomorrow!...

CHAPTER FORTY-SIX

Wearing shorts and white polo, Bubbles climbed the stairs of the train station, as she danced towards me.

"How are you?!" she asked, excited.

She hugged me.

"I'm okay," I said.

I looked down.

"What's the matter, Michael?" she inquired.

"I'm worried about Elizabeth," I said. "I think that it's affecting me."

"You shouldn't let concern for Elizabeth affect your mental health," she responded. "You're in control of your own destiny. You shouldn't let anyone else affect you."

We walked along the streets of downtown Philadelphia. In front of US Treasury Department was a fountain with several sculptures of cats.

"Oooh," she said. "Cats are my favorite. I want a picture of that."

I snapped a photo, then suggested we have something at a local restaurant.

"The last time that I came to Philadelphia," I said, "it was with my Chi Gong group. We were asked to help a woman who had brain cancer. I wonder if she's still alive?"

"It sounds like you have lots of friends, Michael," she responded. "I don't have many close ties. You're the only person that I talk to from Fremont. I don't think that I really like people."

Looking at her across the table, I thought I saw her aura: It was green – the color of healing – and seemed to emanate from a place in her heart...

Returning to walking, we happened upon a secluded park.

"Do you think that there might be muggers?" she asked. "What would they do?"

She looked about like a scared little girl, vulnerable and apprehensive, and I felt a need to get her out of there.

Halfway across a bridge, we stopped, and I sat with my legs dangling over the water's edge.

"I couldn't sit like that," she said. "I'd be afraid."

"What are you afraid of?" I asked.

"I don't know," she said. "I'd just be scared."

Looking down at the river, I depressed the corners of my lips; it was a fifty-foot drop, but I had no fear.

She sat across from me.

"Mike, I am so proud of you," she began. "Somehow you always seem to have it all together. You always seem to know what you want and you achieve it with hard, earnest work. You may not think so, but you seem to know yourself much better than many people I know, including myself. You have been working on you own insight and spirituality for years. It seems that I am just beginning to look inside. I have focused so much on achievement and outward fun that I forgot to look inside. I have recently begun to try to learn more about myself and I can't say I like all that I've learned. But I can say that I am taking steps forward and am developing my own sense of self-achievement and spirituality. I am happy and feel more balanced."

I looked out into the horizon.

"Your research seems to be going well," she continued. "And I think Chi Gong is a good thing, from what I understand. You have always wanted to help others. You are so generous."

I turned to her.

"Bubbles, I'm sorry for pulling away from you all those years ago," I said. "At the time, nothing would keep me from realizing my ambition for science."

"Michael, I have never been angry with you," she said. "I have always enjoyed your company and miss your companionship. I do think of you often, even though I fail to tell you."

I nodded.

"Bubbles, if there's ever a way that I could be of help, I'd like to."

"I'm fine, Michael," she responded. "Really, my life is super. Super-duper. Better than I could have ever imagined…"

"You didn't tell me Bubbles was drop-dead gorgeous!" Liz exclaimed.

I'd been sharing the photos we took in Philadelphia.

"People like her live by a different set of rules than the rest of us!" she asserted.

Depressing my lips, I subtly shook my head.

"She's a person like anyone else," I contended.

"No, Mike!" Liz insisted. "They don't see things the way you and me do. It's different with them. No wonder you've never gotten over her."

She sighed, then came and sat beside me, pressing herself to me and laying her head upon my shoulder.

"In many ways I see your predicament as much worse than my own," she lamented. "The essential difference between us is that you've been in a relationship and left someone you loved. I could never do that. For years you fought to convince yourself that you could live without that love, till finally you had to surrender to those feelings – That love was the most important thing to you…"

Part Five

A year later Bubbles reached out again…

The struggles along the way are only meant to shape you for your purpose.
~Chadwick Boseman…

CHAPTER FORTY-SEVEN

September, 1996

Bubbles informed me that her mother had been in the hospital.

"Mom got sick in Reno," she said. "Combination of heart attack and congestive failure. Her doctors give her a month to live."

"Dad took her home," she continued. "They say that the only thing that could save her is a heart transplant, but that she's too sick for that…"

Liz was in the room and sat attentively listening. Replacing the receiver, I looked at her.

"Bioenergy might help," I said.

Picking up the phone again, I booked a flight, then immediately packed my bags.

"I know you'll help her," Liz said. "Because she loves you..."

CHAPTER FORTY-EIGHT

Ben picked me up at the airport.

Arriving at his home, he held his head in his hands.

"I was out in the heat all day," he said. "I'm having a migraine."

Scanning him energetically, I experienced a sharp pain in my chest.

"Ben, what are you feeling?" I asked.

"I feel like I'm about to explode," he said. "I'm anxious all over."

I followed the feeling. Soon, my body became imbued with a feeling of energy. My chest opened, and I looked out as though connected with all things.

"Thanks, Mike," Ben said. "I feel better."

"Are you still having the headache?" I asked.

"Yeah, I better lay down," he said.

I followed him into the bedroom.

"Ben, there's a technique I know that sometimes helps for headaches," I said. "It's called cranial sacral therapy. Would you be willing to try it?"

"Sure," he responded through gritted teeth.

I felt for his cranial rhythm, and, as I did, I experienced a sensation radiating from my eyes.

"Ben, where is the migraine affecting you?" I asked.

"Around my eyes," he replied. "That's the place where I experience most of my migraines."

Again, I let the feeling flow through me, as I continued to follow his cranial rhythm. Soon, it felt as though my whole skull was breathing in-and-out with his.

"It's getting better," he said. "The pain is going away."

"Rest, Ben," I responded. "I'll check on you later."

From the doorway, I looked back. Surveying my friend's small, slight frame, I wondered how it was that this man – to me – could be such a giant?...

CHAPTER FORTY-NINE

In a foreign land, a James Bond-like figure was being aided by a dark-haired woman in an attempt to acquire hidden secrets. They crept down a flight of stairs, and she disappeared while he searched the room. But shortly after discovering the location of the secrets, he was apprehended by security forces, who bound him, then taunted him about the certain prison and torture and likely a 'bullet to the brain' that he was facing.

Yet he displayed not the slightest hint of doubt or self-pity, accepting his fate with courage and resolve.

And looking on, I wondered, "How does he do that? This is the end. 'King and country.' For what?..."

Awakening from the dream, I shared it with Ben, as he drove me to the bus station.

"For me, it was 'Science and humanity'," I confided. "And now I think, 'Someone else could have come up with the cancer vaccine, and then pursued energy medicine. And I could have clung to her skirt and never gone out and tried to make a difference in the world.'"

Ben nodded and kept his gaze on the road ahead.

"You two were really close," he said. "But you were committed to your work, and she had to move on with her life."

A pained expression swept my features, as I directed my gaze downwards.

"It's OK," he insisted. "We follow our paths and it takes us away from people. Sometimes you ride the wave of life together and sometimes it's just for a short season.

"I think we come into this world with some sort of mission. Not a mission given to us, but a mission of our own choosing – Of our own spirit's choosing. We only have just these glimpses of it, but it drives us, and if we don't follow it, we're miserable.

"Something in your heart, something in your soul, told you way back then that Bubbles was not your objective in this life. And no matter what you could have done, if you stayed with her, it wouldn't have worked, because it wasn't where your spirit was calling you. The human in you wanted it to be, but your spirit was like, 'No.'

"And your path lies ahead of you, not behind you."

I confided that I was still haunted by my feelings for her.

"Mike, if you would have stayed with her, you would have hated yourself," he responded. "Because you would have took your path out of your life. You wouldn't have met Dr. Amin. You wouldn't have met Elizabeth. You wouldn't have gone on and studied bioenergy."

I wouldn't have developed the cancer vaccine, I conceded.

"Right," he responded. "It's like I said, Mike – Your spirit drives you forward towards things. And back then, it was really driving you. You've put in so much work in this life. You have this thirst for learning and seeing things and understanding things and then writing about it. I can't even write a paragraph about my life. So many stories, so many experiences – I can't put it down on paper. But you can. You're able to. You want to do that. This is something that you're inspired to do. This is a path you're inspired to take.

"Don't be sad that this happened. Just realize the truth of it all… That those feelings you had back then we're keeping you from diverting from where you had to go.

"And you might not have known where you needed to go then – But your spirit did. There's a little voice in your head that talks. There's a feeling you get that this is wrong and this is right. It just is.

"And you have to be good to yourself. You have to forgive yourself for taking your path. Because you do burn people along the way sometimes – because they don't always understand – because you can't even explain it yourself. You just know that you've got to do what you've got to do, and you've got to go where you got to go for some reason.

"Mike, you're a really good person. And I mean you're a good person because in your heart you are. You're really tough on yourself, but you really want to help people. You have a noble, courageous heart that wants to go out there and really make a difference.

"And some people might say you couldn't see what was right there in front of you – But it was because your spirit wanted to go somewhere else. And that's good. That's fine. That's who you are. That's who in all your uniqueness you are. And there's absolutely nothing wrong with that."

Still, I felt a sense of doubt: The pain in my chest that I experienced those years without her – I hadn't followed my heart.

"I don't know that you didn't follow your real heart back then," he asserted. "I spent many, many nights up late on the phone with you during your time with Bubbles. And did you love her? Yes, you did. Do you still love her? Yes, you do. But it doesn't mean that it would have worked.

"What would you have been willing to give up for a life with her? Would you have been willing to say, 'OK, no, I'm not going to go to the NIH. I'm just going to get my degree here at Fremont, and open up a practice, and we're going to stay together, and our dream is going to have a family, and I'll be a practicing doctor'?"

No, I wanted to pursue cancer research. I wanted to parlay what I'd achieved in college and accomplish even more. Now I wanted to do the same with bioenergy.

"OK, then you weren't going to give that up for her. And that was like a definite.

"And imagine if you did give that up. How miserable do you think you would have been?"

Yes, the me from then would have been miserable. I would have felt like I'd failed in my life's mission. In turn, I likely would have turned that on her.

"Right, so your spirit protected you – And her! Protected her from that miserable life together – Because you loved her!

"So, in retrospect, say, 'Thank you, spirit. Thank you for making me do what I needed then.'

"And believe in who you are. Believe in your spirit. It's a good spirit. It's a really vibrant, good, honest spirit.

"And there are a lot of things you don't see in life – But that's because it's unimportant for your journey.

"Maybe your journey is going to help you just wander and go from one to another, place to place? Who knows? I just watched a show about this young woman who's in Africa now. She's doing African healing and finding your spirit through dance – to the drumbeat and ancient, ancient ways of connecting with who you are and everything. There's so much magic out there like that. There are so many amazing experiences out there.

"And you're doing the hard work to qualify you to go anywhere you want. When you're done with medical school, you can sign up for Doctors without Borders and go and explore all kinds of stuff if you wanted to. You can go work in Third World countries, and go dance with the natives, and learn their magic and their healing. You learned bioenergy. What kind of energies are there in sound? What kind of energies are there in dancing in a drum circle? How healing to a soul is that?..."

At the bus terminal Ben told me he couldn't stay to see me off.

"Because I need to get some things done at home before my next shift," he explained.

Returning to his truck, he turned.

"Take care of yourself," he said. "And respect who you are. Respect your choices. And respect your spirit! And listen to your spirit – What it's telling you. It's been right all along..."

CHAPTER FIFTY

Golden reeds stretched along the grapevine; canyons that were burnt and charred, with clay savannahs of brilliant red with blue undertones.

Stepping off the bus at the Fresno depot, Mr. Vasquez was there to greet me.

"Hey, there," he said.

I followed him to the truck.

"The doctors give her a ten percent chance of surviving the surgery," he explained. "We had to go with it, though, because there's no other option..."

Arriving at the house, family from across the state filled the living room. Crossing the threshold into the darkened bedroom, Mrs. Vasquez looked at me.

"Hi Mike," she said, weakly.

"Hi Mrs. Vasquez."

I knelt at the side of the bed and kissed her. Her skin appeared ashen gray, but her cheek felt soft like Bubbles'.

"Bubbles told me you weren't feeling well," I said. "I wanted to come and try to help you with some alternative treatments I've been learning. After I saw you last, I went back to DC and found a doctor who got my leg better. He did something called bioenergy, and when it helped me, I asked if he would teach me the technique?"

She listened, intently.

"Bioenergy is built on the concept that there's an energetic circulation to the body – One that science hasn't exactly defined yet. When the body sustains injury or trauma, blockages in the energetic circulation develop. The source of this trauma can be emotional, physical, and, left untreated, can manifest in physical disease. Bioenergy can remove those blockages and restore the body's normal flow of energy. Now, I'd like to try to help you with the technique. Would you be willing to try it?"

"Yes, I believe in all that," she responded.

"Great, let me start by trying to demonstrate it to you," I said. "Not too long ago, my friend, Elizabeth, discovered I emanate an energy that people can feel. I use it now to make bioenergy more tangible to people and take away some of the mystery. Would you be willing to try it?"

"Yes," she said in a hoarse voice.

I held out my hand.

"Put your hand over my hand," I said. "Try not to touch it – just keep it about six or so inches above my hand, and see if you don't feel something rising from a point in my hand."

She moved her hand back and forth above mine.

"I feel heat," she said. "It feels like it's coming from the middle of your hand."

"Great," I responded. "Let's see how far your sensitivity for this goes... Move your hand up and down my forearm and see if you can't find another point."

She guided her hand over my arm.

"I feel it over the middle of your arm."

"Wonderful," I said. "Now that you have a feeling for what bioenergy it, I'm going to move my hand over your body, looking for points of energy coming from you, not unlike the ones you felt over my hand."

Scanning her energetically, she closed her eyes. Quickly, I found an energy disturbance emanating from her abdomen; allowing it to move my hand in the direction it was flowing, I found it radiated to her throat.

"Perhaps it's because I have been so sick to my stomach?" she commented.

Continuing to move with the energy, my hands radiated with more energy than they ever had before.

Then, to my complete surprise, my hands became utterly chilled!

"My hands are cold," I said, startled. "I mean, they're ice cold."

Looking upon them, it felt like they weren't my own – but, rather, belonged on someone decades older who suffered from severe rheumatism.

"Perhaps you've used up all your energy on me?" Mrs. Vasquez responded, half-jokingly.

She smiled and opened her eyes.

"Thank you, Mike," she said. "You've made me feel more relaxed…"

Stepping outside I strolled through the neighborhood. Passing a familiar farm house, a dry wind blew and the frozen feeling in my palms dissipated, recharged by the sun...

Returning to the house, the guests and family members were now collected in the yard. A sweet little boy named Jojo played along the fence; when he fell and pinioned himself on a barbed edge, he began to cry.

Mr. Vasquez reacted angrily and gripped the small boy.

"You stop crying!" he demanded. "Or I'll give you something to cry about! I take off this belt and give you a whipping!"

The boy's parents whisked him away.

"Boys shouldn't cry," Mr. Vasquez continued. "Nothing to cry about. He shouldn't cry anymore. You have to control them, or else they control you."

He turned to me.

"What do you think?" he asked.

I shrugged.

"Hurt feelings hurt," I responded, softly.

As the guests prepared to leave, Mr. Vasquez gave the boy five dollars.

"You take this," he said…

CHAPTER FIFTY-ONE

In the evening the Vasquez's eldest daughter, Sylvia, arrived at the house. A registered nurse, she conveyed more specifics about her mother's condition.

"The doctors say she's had two undiagnosed heart attacks, a leaky mitral valve and congestive heart failure," she said. "They feel that her heart is so bad that there isn't time to let the ulcers heal before performing surgery."

I listened.

"What do you think, Mike?" she asked.

"I can't help but think that she might do better if we help heal the ulcers before she goes in for the surgery," I responded.

Mr. Vasquez had been sitting with us at the table; he shook his head, disgusted, then got up and marched off.

As Silvia departed, I stood by the doorway I admired the framed photographs on the wall. In one of the photos Little Davy was perched on my shoulders, with Nellie and Bubbles standing on either side of us, as we smiled in our bathing suits.

Our day at the waterslides, I thought.

Davy and Melissa have moved to North Carolina after she'd taken another husband, and I wondered how they were?

Just then, there was the sound of a thump, followed by a loud crash and a woman's voice calling out. Hurrying down the hall I found Mrs. Vasquez laying in a heap on the bathroom floor, Mr. Vasquez already assisting her.

"You see!" he shouted. "She needs surgery!"

We led her to the bedroom.

"I hurt my neck," Mrs. Vasquez cried.

"Can I help?" I asked.

"No!" Mr. Vasquez insisted. "You let her sleep now!"

But she continued to whimper, and Mr. Vasquez appeared at my room.

"You see what you can do," he said.

Cradling her head and neck in my hands, I searched for her cranial rhythm. The rhythm was discordant, as though there were no rhythm at all – only trembling. Finally, I managed to locate something that felt like a pulse and followed it as it migrated from place to place. Following it, I noted that Mrs. Vasquez's muscles were supple and her hair was thin, and again I was reminded of Bubbles.

It was impossible not to notice that her muscles were similar to Bubble's, as was her soft hair. As time passed, the rhythm became more definite. Then, still points occurred, in which the rhythm stopped entirely, signaling her body was healing and undergoing self-repair.

But it was a long process, and, as the minutes ticked by, the sun came out and dusk transformed to dawn, and having nothing else to do, I let my eyes wander about the room until they settled upon the nightstand, on which was a number of children photographs. Studying the faces of the children, I was able to make out the features of each of the Vasquez's spirited daughters – Until I came upon one photo of a little girl who (for the life of me!) I couldn't recognize. And it was odd because her photo was especially encased within a gilded silver frame!

Appearing shy and downtrodden, the girl's hurt eyes were full of tears that looked away from the camera's lens; and a little bird was perched upon her shoulder that seemed to be trying to sing her grief away.

This must be a distant cousin, I speculated, who lives in some impoverished place in Mexico.

Still, I couldn't take my eyes off her, as I experienced some deep-seated feeling of empathy and kinship for her.

Then, I noticed that Mrs. Vasquez was smiling and staring at the photo with me.

"Who is she?" I inquired gently.

"Bubbles," Mrs. Vasquez answered with affection.

"'Bubbles'?!" I responded, astonished.

Mrs. Vasquez nodded, and I looked back at the photo, smiling.

We're cut from the same cloth, I thought…

CHAPTER FIFTY-TWO

In the morning Mrs. Vasquez's abdominal pains were getting in the way of her having breakfast.

"The cranial therapy that you did last night seemed to help," she said. "Why don't you do that now?"

Following her cranial rhythm again, a pulse developed, until, finally, her whole head radiated with energy.

"It's not a still point," I told her, "but rather an indication that the body's received as much as it can for the time being."

"Maybe you can do acupressure?" she asked. "We have that device that the chiropractor uses – an activator. It's in the dresser drawer."

I stimulated points along the stomach meridian. They were sore and tender, and, ultimately, provoked dry heaves.

Her system is so depleted, I concluded. I have to find a gentler way.

"Mrs. Vasquez, I'm going to try to initiate you into the practice of bioenergy and Chi Gong," I said. "Try to put yourself into a meditative state. I'm going to talk you through several of the concepts related to its practice. Let's see if we can tap into your body's innate intelligence and facilitate its ability to heal.

"Energy Medicine approaches like Chi Gong and bioenergy are built on the concept that there's a universal life force that runs through our lives. When the body sustains injury or trauma, we may become disconnected from this life force. Chi Gong and bioenergy are meant to bring it back in line.

"I know how devoutly religious you are – Chi Gong isn't all that different from prayer. Let your body enter a meditative state – one in which the healing energy of the universe can come in and give the body what it needs. Part of this is tapping into the innate intelligence of the body. The body knows how to undo all injuries and traumas. The trick is to reconcile the body with the source of injury so that this intelligence will be released. I think that it will come to you easily. Just get into a meditative state, so that the body can become acquainted with the sources of trauma and let those imbalances go.

"I think that a lot of this involves the invisible power of love. It seems that when you think of another with deep appreciation, it imbues you with a certain reverence for the person and permits you to channel and tap into the universal life force, so to facilitate healing. The universal life force is what I regard as a central blueprint for all life, that contains within it the capacity to correct all imbalances and undue any strain, trauma or injury. I don't regard this as different from any form of prayer or worship – I simply recognize its medical utility. And it's taught me that only through compassion can we derive a greater understanding of the world around us.

"I believe that all things are connected and came from the same place. That doesn't mean that we're all the same – we're different. But, as parts of the whole, it becomes crucial that we realize our true selves and live to our potential, so to contribute our best to the universe.

"In bioenergy one person helps another by correcting an imbalance, thereby restoring harmony. Whether the affected part be a limb or an organ that is suffering, it holds the body in a disorganized state of chaotic impulses. Think about it like a muscle spasm: In a spasm, usually the person's taxed the muscle beyond its limit – to the point it can no longer function as before and work the way it used to. The rest of the body is frustrated, because its usual balance is out of whack and now the entire body is unstable. Wild attempts to restore order only exacerbate the condition – and, really, that's not what the muscle needs, and only contributes to a worsening of the injury.

In this case the muscle needs to tap into its former self, so that it can return to its normal state. In Chi Gong and bioenergy a communication is affected between the strained part of the body and the universal life force, so that its original blueprint can be availed and the muscle relearn its normal function. When that happens, the function is restored and the system goes back to working. It's not unlike what happens when an electric shock is delivered to a heart in cardiac arrest. Sudden death usually involves ventricular fibrillation – a chaotic set of signals going off everywhere in the heart. When the person's heart is defibrillated, the resultant depolarization clears those random impulses, and the heart 'remembers' its original rhythm and restores its normal beat.

"I believe there is nothing that is insignificant: Everything you do effects something else until that action is felt throughout the world. Getting in harmony with your mind and body is a wholly individual endeavor – Everyone has different wants and needs. Permit your body-mind to connect with this universal theme and harmony can be restored.

"I believe that at the heart of everything is love. Love is the only energy I know of that is not only recyclable, but self-propagating. Because when you love someone, there's the potential for it to grow and snowball and spread one to another. Touch love, and life is rekindled; let love lay dormant, you forget.

"Because love is life's essence – Love for a moment can last a lifetime – Can last for all time…"

Slowly, Mrs. Vasquez opened her eyes.

"While you were talking," she said, "it was as though your words were coming from far way, and I could hear them echo deep in my mind."

Then, she nodded.

"You must always love," she said. "That is for sure true. You must always love..."

CHAPTER FIFTY-THREE

In the Vasquez's garden, my brother, Joe, was planting roses. I'd been sleeping within the house till awoken by a flash of light. Bubbles was at the gate, light shining all around her. As I opened the door to meet her, Mr. and Mrs. Vasquez standing just behind. Coming near, a fair-skinned version of Bubbles stood before me. Wanting her to greet her parents first, I moved to step aside. She smiled and pulled me close.

"It figures!" she exclaimed. "Didn't miss a beat!"

This Bubbles was Elizabeth...

Awakening from the dream, I heard the sound of chirping of crickets, then the swinging of a gate and the revving of a muffled engine.

Opening the backdoor, I strained to focus my eyesight, as I peered into the headlights coming out from the night. Nellie was waving to me enthusiastically from the car, and Bubbles was standing to the side holding open the gate, wearing the same set of sweat clothes that she'd worn in college.

It was 3 AM – Yet upon entering the house, 'the girls' bounced about as though nothing out of the ordinary had happened, and nothing at all had changed; and in their mother's room, they chided her for being sick.

At the doorway I stood looking on. Bubbles turned to me.

"I'm so happy you're here, Michael..."

CHAPTER FIFTY-FOUR

In the morning, we learned that Sylvia had been admitted to the hospital following a bout of breathing difficulty due to her asthma. Bubbles, Nellie and I piled into their convertible to visit her.

"The three of us in a crowded car," I thought. "Like the first time I came here, and Nellie was learning how to drive..."

I bought flowers and we brought them to Sylvia. Sylvia's husband, Miguel, stood in a corner.

"I had to get off work to come here," he said. "My boss calls us 'wet backs' because he says we all came over the Rio Grande. I tell him that he came over the ocean – I call him 'salt back'..."

Leaving the hospital, we went to a Christian book store, where Bubbles purchased a Bible study guide. After, we ate at a Japanese restaurant. I told the girls their father was under a lot of strain and recounted the day before when he'd got upset with Jojo.

"He's a violent person," Bubbles said. "I'm a violent person, too."

I looked at her.

"You're violent?" I asked.

"Uh-huh."

"I've never seen you violent," I said.

"Well, I am," she responded.

"When have you been violent?"

"I'm violent a lot," she replied. "Nellie's seen me get violent."

"Well, I've seen you get into fights with Tim and stuff," Nellie commented, "but I don't know if I'd say you're a violent person…"

Leaving the restaurant, we drove to a church. Bubbles turned to me.

"I started being violent when I left Fremont and went off to Lake Tahoe," she said. "I'd been drinking a lot. I guess that I just needed to grow up…"

In the church the musicians played rock-and-roll accompaniments to different hymns. Standing beside Bubbles, I appreciated the child-like quality of her voice.

The church leader ascended the pulpit. He was a vigorous man in apparent good health.

"We are going to restore religious order to our schools," he asserted. "We are taking aim at the homosexuals in our classrooms and remove them. We're going to ferret out teachers in favor of abortion and target them, too..."

Outside, the desert wind blew through my clothes.

"What did you think of the service?" Bubbles inquired.

"I thought that it was flawed by intolerance," I responded.

She hesitated.

"I agree," she said. "Do you go to temple anymore?"

"No," I responded.

"Then, how are you cultivating your spirituality, Michael?" she asked.

"Mostly in the work I do," I said. "When I do bioenergy, I feel connected at every moment."

"To what?" she asked.

"To God," I answered.

She looked at me.

"That's good," she said…

CHAPTER FIFTY-FIVE

Arriving back at the house Mr. Vasquez was angry.

"I tried to find you!" he rumbled. "I wanted you to give blood for your mother today."

The girls walked past as though unconcerned.

"We can go tomorrow," I said.

Shaking his head, he scuffed off.

Returning to the living room, Bubbles sat in the recliner across from me.

"Michael, show me this work you do," Bubbles said.

I held out my hand to demonstrate bioenergy for her.

"I don't feel it," she responded.

Indeed, I wasn't feeling anything either. With anyone else, I at least felt energy projecting from my hand, but, now, with her, there was just dead air that I was experiencing. Like I'd put my hand through some veil, for which on the other side was a blackhole where everything was got sucked out and swallowed forever. Just this utter nothingness and void between us. Nothing but empty space.

"That's funny," I responded. "Most people do."

She was the first person for whom I had not been able to make bioenergy tangible.

I looked at her confused, as a feeling of sadness gripped me: I knew that no matter what I did, I'd never be able to really share this with her. Not in any meaningful way. It would always just be words to her and nothing she'd ever be able perceive and have a feeling for.

This person with whom I'd wanted to share everything. With whom I had shared everything.

My thoughts drifted to college on that day she couldn't bear when I wouldn't share something with her. It was the week that Professor Hope got upset with me because I'd created a computer problem. He cursed at me, and I left the laboratory feeling crushed; I thought my dreams of a career in science were over, and felt so ashamed of what I'd done that I remained hunkered in my room, feeling like I'd never recover.

For days, the only time I left was for meals at the Dining Commons. That's where Bubbles saw me from clear across the Commons, sitting at a small corner table with hundreds of students in between.

"I think there's something's wrong with Mike," she told Tim.

"Oh, he's probably just mad at me again," Tim responded, laughing.

"No, I think it's different," she replied...

She was sitting at the end of our usual dormitory floor communal table, as I was hastily making my way past, then looked up at me and smiled.

"How you doin', Mikey?" she asked.

I stopped.

"Not good, Bubbles," I said, my head bowed and shaking.

"What's the matter?"

I hesitated.

"I can't tell you," I said, completely uncharacteristic and contrary to any way I'd ever been before.

Departing, I went back to the solitary confines of my room.

Not long after, though, to my surprise, there was a knock at my door, and when I opened it, Bubbles walked past me, entering the interior as though in a daze.

"Mike, you've got to tell me what's the matter," she insisted, pacing back and forth and almost stomping her feet. "You've got to tell me what you're feeling."

I nodded and said I'd tell her if she'd accompany across the street to the market. She came and we talked, and then she came back to my room and sat on the bed and comforted me, until she'd chased all my dark clouds away and replaced them with skies of blue.

Then, she held me and told me that she loved me, and kissed me on the neck, so to send chills all through me.

I was hoping it would always be that way between us. That I'd be able to share everything with her for a lifetime.

I'll never be able to share bioenergy with her? I thought, despairing. How could this be? Why would this be?...

Scanning her, energy radiated from her chest and moved in the direction of her sinuses.

"What do you think it means?" she asked.

"I'm not sure," I answered. "Are you having problems with allergies?"

"I've been suffering from asthma," she said. "Usually, I have trouble breathing at night."

Nodding, I intentionally held back from saying that, according to Traditional Chinese Medicine, imbalances of the lungs suggest unresolved sadness…

Making our beds – mine on the floor and hers on the couch – she turned to me.

"What other kinds of things do you do, Michael?" she asked.

I shrugged.

"Cranial sacral therapy," I responded. "If you like, I can show you."

Positioning myself behind her, I placed my hands on either side of her head and followed her cranial rhythm, which was even and regular.

"Bubbles, put your hands on mine and see if you don't perceive the movement," I said.

She lifted her hands and gently pressed them to mine.

"Do you feel that?" I asked.

"Yes, I do," she said.

Just then, the rhythm stopped, and in its place came a chaotic release of spasms that felt like a lot of undulating worms under my fingers.

Still point, I thought. But that's something that happens when there's a problem? There's no problem here.

Slowly, her cranial rhythm returned.

"I wonder why that happened?" I commented, confounded.

"I'd been having a stomachache before we started," she responded. "I don't have a stomachache anymore."

She smiled.

"I'm proud of you, Michael," she said. "I'm proud of what you're doing with your life."

Returning to my place on the floor, I sat with my head bent.

"I still feel like I'm waiting for something," I lamented. "For a moment when I'll declare, 'This is what I love – who I am – what gives my life its greatest meaning.' I just wish I could live with my mistakes."

She looked away, then got under the covers.

"Thank you for the day, Michael," she said...

CHAPTER FIFTY-SIX

In the middle of the night, I awoke to the sound of coughing. Standing in the doorway of the unlit room, Bubbles was gasping for breath, as she labored to take in the cool air of the night.

"I'll be alright," she insisted.

Leading her back to the couch, I sat with her until the attack passed. In between, we talked about the last days, and I reiterated my surprise at what happened between Mr. Vasquez and Jojo.

"He treated us that way," she responded. "If I ever cried, he'd thump on my chest and tell me to stop it. I had to suck in my tears."

I shook my head, disbelieving.

"I forgive him, though," she added. "I forgive my father."

Just then, Mr. Vasquez stormed into the room.

"Your talking is keeping your mother awake!" he shouted at Bubbles. "By leaving the door opened, you've made the house all cold! You go in the room now and sleep with your mother!..."

CHAPTER FIFTY-SEVEN

In the early morning Mr. Vasquez stood at the stove preparing breakfast. I sat with him, and he put a plate of potatoes and tortillas on the table for me.

"You need to comb your hair," he said. "There's a right way of doing something and a wrong way, and you want to do things the right way."

I smiled.

"I don't think that it's always that clear cut, Mr. Vasquez," I responded.

"There is always – always – a right and wrong way of doing something," he insisted.

"And what about the shades of gray in between?" I asked. "That cloud and obscure the determination of right and wrong?"

"I'll give you an example," I continued. "While I was in South Africa, I gave a woman a ride. She was colored – of mixed decent – and, for that reason, had a special place in the Apartheid system. She said that she'd been reasonably content with her place in society; but when I condemned Apartheid as an evil, discriminatory state-sponsored system, not different from the Nuremberg laws that Hitler imposed upon the Jews before annihilating them in the Holocaust, it seemed my words had the effect of 'destabilizing' her – replacing her fragile faith with a cruel reality."

He'd been listening like the gambler he'd told me his father was – with steely eyes that concealed his hand.

"Well, did you do right or wrong?" he asked.

I hesitated, depressing the corners of my mouth.

"I'd spoke the truth, but what I did was wrong," I responded. "If I could do it over again, I would have done it differently."

"That's where you're wrong," he asserted. "You'd done right in the first place. Even if that woman went and blew her brains out after what you told her, you still did right to tell her how things were, because it might have helped her to better her life. You done right and shouldn't think about what she's gonna do – Because you done right."

"You have something wrong with your thinking," he asserted. "Have to change it now or else it get you into trouble."

He shook his head.

"Why you do different with that woman in the car?" he demanded. "Why?"

"Because I scared her," I responded, "and I care about her feelings."

"That's no good!" he countered. "You say that, you say that I don't care about people."

"That's not what I said, Mr. Vasquez..."

"No," he interrupted, "but that's what you're saying."

He got up.

"You have to get straight in your mind," he insisted. "No, 'What about this?' or 'What about that?' Right or wrong – that's it. We shouldn't even be talking about this…"

CHAPTER FIFTY-EIGHT

After donating blood at the Red Cross, Nellie, Bubbles and I drove to the County Fair. There, the Vasquez's middle daughter, Edie, was waiting with her children – an agreeable infant named Christopher and a three-year old named Zackary.

As we ventured into the fairgrounds, Zackary lagged behind. Watching him, he seemed painfully shy and sensitive, and I wondered how I or anyone could reach him to show him that the world wasn't such a fearful place and he didn't have to be so guarded and afraid?

Just then, Bubbles swooped in and hoisted the boy upon her shoulders and led him from ride-to-ride. Filled with excitement, his laughter filled the air, and watching them, I wondered that perhaps she might save him from a fear-filled childhood and infuse him with that spark of joy?

And following them through the fairgrounds, my mind wandered – confusing Liz with Bubbles, me with Zackary, as feelings and people flowed one into another…

CHAPTER FIFTY-NINE

As I lay in the living room floor, a memory surfaced of being in a park with my father and his girlfriend, Nikki. I was perhaps Zachary's age, and trailing behind Nikki and my father; when I indicated I was tired, Nikki reacted by hoisting me up and over her shoulder (like I were some sack of grain) and carrying me away.

Watching helpless as the ground moved beneath me and feeling terrified of falling, I called out.

"Let me down!" I cried. "Let me down!..."

Sitting up now, I stared in Bubbles' direction on the couch.

Bubbles was tall the way Nikki was. I considered how I used to imagine that she was to me what Nikki was to my father.

But, now, I realized she was the very antithesis of Nikki...

Then, I became cognizant of a feeling of energy everywhere about me, as though it filled the room; and rising to my feet, I was gripped by the odd sensation of walking as though it were no effort at all!

What's happened?! I thought. It was like something shifted in my body and everything just fell into place, so that I was walking for the first time in my life!

I wanted to cry out, "Hey, everyone! Look!"

But the house was still and quiet, and gazing again at Bubbles, I saw she was wearing her familiar blindfold and peacefully sleeping, as I imagined the others were, too.

Creeping to the doorway, I stepped outside and continued to walk as though gliding! Each step requiring no effort at all!

What's happened? I continued thinking. What's responsible for this change?

Weightlessly, I moved about the neighborhood, light as a feather – Requiring no effort at all for the purpose of self-propulsion – Just me moving my body in any direction I wished – Turning at will with no effort whatsoever! It was like being totally at home in my body – Totally in control. At the age of 32, I felt balanced for the first time.

It felt like some 'shift' had happened – Some alignment that perhaps began at the base of my spine and then extended and expanded upwards, so that everything in my body was integrated from top to bottom and bottom to top.

It was as though some major kink in my system had been extricated from my body. There was nothing pulled out a place. No impediment to movement. Nothing for which I had to compensate. Nothing compromised. My body was no longer this assembly of Tetris pieces; but, rather, one integrated unit.

That's what it takes, I thought. Not trying to fit any pieces together; instead, all the pieces are where they should be from the start and working together as they were meant to. They have self-corrected, so there's no forces getting in the way.

All of my muscles felt relaxed... Like they were in neutral, with no hint of being tight and tense and pulled apart this way in that way. Just everything aligned.

The feeling bordered on 'possession' – Like I couldn't recognize my own body – So different was its movement than anything I'd ever experienced before!

It's probably what people take for granted their whole lives, I realized. The simple act of walking. This is how athletes can run. This is how dancers can dance. It's because they have this stability, coupled with an integrated system that facilitates smooth, effortless movement. No wonder everywhere people are walking full of confidence. Being able to move like this instills confidence. How can you feel anything but confident when your every step is so solid? It makes no sense to feel anything else. Being this sure footed. The ground is solidly underneath you, instead of it feeling like it's shifting all around you, so that you're about to tip over at every moment.

I recalled my Physical Medicine & Rehabilitation instructor from medical school and how he'd refer to walking as a series of "well-coordinated falls."

But there's no falling here. I couldn't fall if I tried at this moment. Not as long as there was solid ground beneath me. This movement is too fluid. I'm too connected.

Indeed, it didn't feel as though I needed to hold myself up at all; I might as well be being pulled up to the heavens themselves. There's no effort in this walking whatsoever. Only this fluid movement. I can go any way I want without any effort. None. Just my body mind taking me to where I want to go, effortlessly. I require nothing and no one. Where there's no effort required, I feel no need to be held up and therefore no need to be held. I am completely autonomous. Totally content within myself.

Even more than feeling centered, it felt as though I was in no place – Because I might as well had not even be there. I didn't 'feel' myself. That's how in sync and connected I was with everything around me – the entire universe. I might as well be invisible, composed entirely of atoms and empty space. Indeed, existing only in open space. I felt none of my mass; instead, I felt hollow. Like I am nothing except this conscious entity that only exists in space.

The way I felt, someone could put their hand right through me. The wind could blow through me. Water could rush through me. I am just this entity in space. I am no longer this glob of traumatized tissues trying to make its way through the physical plane. This is what the philosophers talk about when they described being nothing and everything at the same time. I am nothing. I feel nothing. But at the same time, I am connected to everything everywhere.

I experienced a sense of euphoria: Until this moment I had never grasped how extraordinary a physical creation humans really were!

Because there was no way I could have – It had no understanding. It had been taken away from me from an early age.

I wondered how this could have happened? Had I tapped into something or someone who'd served as a channel of healing for me?

My thoughts drifted to Nellie: At the County Fair I'd watched her as she pushed her nephew's stroller, admiring her movement.

Like some well-oiled and integrated machine, I'd thought.

Now, I was exhibiting this perfect walking. I was the well-oiled, integrated machine!...

Before going to the Fair, I'd performed bioenergy with Nellie. Scanning her energetically, I'd perceived a strong feeling of energy emanating from the base of her spine.

"This is the area of the first chakra," I'd commented. "It's supposed to be associated with our ability to reproduce, stand and survive in the world. Any thought as to why I might be perceiving something there?"

"I recently had a gynecologic procedure done," she answered. "I'm trying to get pregnant."

I nodded.

"Well, whatever this is," I responded, "I hope the bioenergy helped…"

Now, I wondered that I'd been the main beneficiary of that exchange?

"I'll never understand how you can have these giant feet and yet you're totally off balance," Liz had said of me.

My feet had been a source of shame. I'd inherited flatfeet from my mother's side of the family. When my father spoke disparagingly of them, I resorted to walking on their sides and concocting every manner of homemade prosthetic to torture myself into achieving something akin to arches.

Nellie, meanwhile, was amazing on her feet! An elite high school athlete, she excelled in all manner of sport.

I remembered the times that Little Davy and me would challenge her to a game of basketball, and despite the two of us teamed against her, she'd still smoke us – knocking down jump shot after jump shot, launching swoosh after swoosh. She was that athletic. That grounded. That precise. That perfect a physical specimen.

Had that short bioenergy session with her gifted me some energy of healing at the first chakra? Helped ground me? Helped align me? Was that what this is? Some transposition? Whereby her energy was transposed upon me to facilitate healing for me to live and experience?

I wondered if I had ever experienced anything like this before?

I hadn't physically felt anything like this before, I decided. But the experience of someone's energy transposed upon myself?... Yes, interlink. That I had experienced before.

However, this time, instead of perceiving someone's hurt, it was like I had tapped into her health?!...

"Yes, a transfer occurred," Ben agreed when I shared the experience with him. "Maybe it's because you're working on Bubble's mom, so that you're connected to them – and not connected to your own body? So that your body has got to function normally, without the control your will puts on it?..."

CHAPTER SIXTY

Climbing a jagged mountain, I labored to propel myself upwards. At the top I saw Bubbles standing and following the progress of my ascent.

Then, rounding the final turn between two spires at the peak, I closed my eyes, leaned back and let myself go falling back to the earth without fear...

Awakening from the dream, my chest felt full and strong; the familiar pains in my body were nowhere to be found and the injured muscles of my leg were unwinding.

Whatever it is, I thought, letting go is the right thing to do.

I laid back.

'Without the control of my will'? I thought, considering my friend's words. The one time in my life that I didn't apply my 'will' and allowed myself to 'let myself go' was that summer with Bubbles – when I'd just let things happen. Is that what I'm experiencing now? Am I immersing myself in that energy again?

Then, suddenly, I felt this convulsive 'pop' happen inside me! It was an earthquake at my midback! And it occurred with such force, so to bounce me off the ground!

What was that?! I thought. It was like a couple continents had moved over each other inside me, creating a seismic shift within that was literally off the Richter scale!

Just then, the muscles in my back and shoulders began to fire spontaneously, as though a cascade of trigger point releases had been set off all at once! The muscle twitches were so intense and rapid that they sent me flopping on the floor like a proverbial fish out of water!

This is like what happened to Liz, I reasoned. When she underwent that neuromuscular reeducation. But what is it from? I didn't fall off a rocking horse.

The focal point was the center of my midback. That's where that Gilmore boy hit me when he knocked the wind out of me – knocked the life out of me. This is getting at what happened all those years ago in childhood. It's what's been physically holding me up. It's probably why I couldn't get better when I had the leg injury. It's probably the key lesion. And for whatever reason my body has decided that now is the time to tackle it!

The releases from the back were generating so much movement at the shoulders that they were sending my arms literally 'flapping', as if I had wings!

Then, the releases went to the muscles of my midback; the contractions extended to all the way around my ribcage, so to open all this room in my chest for my heart; and had the feeling of an osteopathic manipulation, such that my breathing was so much more relaxed after.

But why are these releases so 'explosive'? I thought. Some felt like they were lifting me a foot in the air from where I lay? As though they were meant to throw off all the books and clipboards that I'd kept close to my chest and were suffocating my heart?

My thoughts drifted to Don Ko and something he'd told us during a Chi Gong practice.

"If you don't take care of trauma when it happens," he'd said, "if you don't massage the muscles after they've been injured, then they become like cement."

"And once they're 'cemented'," he continued, "the only way you can break out of it is to break the cement."

That's what these muscular releases felt like – Like they were breaking through some cement that had been laid down in them.

What if the 'cement' were overlapping muscle myofibrils – acting and myosin – which had been locked into spasm and hinged so tightly that they can't come apart without an explosive amount of energy to unhinge them? Were these muscular releases actually 'jump signs' in trigger points to unlock them? So that afterwards those filaments would slide by each other and let the muscle relax? Are the trigger points the cement and the muscular 'explosions' basically large jump signs enabling the actin and myosin filaments to come together enough to unhinge and disengage before releasing relaxing?

Furthermore, despite the explosiveness of the releases, the process was entirely without pain!

It's because there's no inflammation there, I reasoned. What's being acted on are the manifestations of injuries from long ago. Just tired old cement waiting to be broken up to reveal the healed tissue underneath. Hence, there's no inflammation to create pain. It's purely well-orchestrated releases that the body is conducting and overseeing. So, as startling as the jump signs were, there was no pain involved.

In an effort to keep my thoughts from getting in the way of what my body was doing, I decided to distract myself by immersing myself in a pleasant memory. I remembered the photograph of Bubbles and me and Little Davy and Nellie on the wall; the room was too dark to see it clearly, so I closed my eyes and brought the image of another photo (Bubbles and I holding each other overlooking Lake Tahoe) into my head.

But the deeper I immersed myself in the thought of being with her then, the slower the releases became until they stopped altogether!

Now that's interesting, I thought. My body deeply relaxed, it was as though I'd released some flood of endorphins within.

Bubbles is medicine to me, I realized. Like a narcotic meant to sooth the hurt child within.

But the kind of medicine she represents isn't what I need now. I need a different kind of healing – If I'm ever to recover, I've got to rid myself of these trigger points that are keeping me locked up so I can't breathe and live fully; I need to get at the source of trauma and not just suppress the symptoms!

Letting go of the image, my body went back to the muscular releases that it had been undergoing before. Each successive series of releases opened my chest further, so to facilitate my ability to breathe.

Rising to my feet, I experienced a fullness in my chest and ease of breathing that I never thought possible? It was as if some physical obstruction in the hinge of the gate to my heart (that had always been there and kept stuck and incapable of working) had been removed, so to be open and unlatched?...

CHAPTER SIXTY-ONE

From the other rooms I heard the sound of stirrings. I'd returned to sleep and it was the wee hours of the morning and entirely dark inside the house. Muffled voices came from the interior of the home, punctuated by the faint bark of orders.

Then, I perceived movement in the hallway, and, focusing my gaze, made out the image of Mr. Vasquez, then Mrs. Vasquez at his side.

Instantly rising to my feet, I got to the other side of Mrs. Vasquez and assisted her through the front door, to the outside and into the truck.

"Mike, you come with us!" Mr. Vasquez commanded. "Bubbles and Nellie – you meet us at the hospital..."

Arriving at the hospital, it was still pitch black outside. Having sped ahead, Mr. Vasquez had lost Bubbles and Nellie some miles back. After assisting Mrs. Vasquez to the hospital entrance, I went back to the parking lot to find the sisters.

Recognizing their car as they pulled into the lot, I waved and went running in their direction. Reaching them, Bubbles leapt from the car and hugged me...

Inside, Mrs. Vasquez had already been taken back to the operating theater. Sitting at the edge of a chair I anxiously awaited news of the surgery. Bubbles sat across from me, smiling full and radiant...

Finally, the thoracic surgeon appeared and declared the surgery a success.

Mr. Vasquez smiled, triumphant.

"I give all the credit to the surgeons," he said. "They're the ones who fixed it. They took care of the problem…"

In the recovery room Mrs. Vasquez was wrapped in white blankets, the color having already returned to her complexion.

"Thank you, Mike," she said. "Your treatments really helped."

Scanning over the surgical site, all I experienced was a feeling of peace.

Yes, I thought. My work is done here...

Preparing to depart, Bubbles kissed me.

"I love you, Michael."

Then, she and Nellie piled into the convertible and peeled off...

Driving me to the bus terminal, Mr. Vasquez was mostly silent.

"When you're a famous doctor," he began, as the Greyhound station came into view, "and an older couple comes to you asking for medicine, and you take care of them, and you charge them a dollar, and all they have is 90 cents, you give 'em a break. Give 'em a break…"

Taking a seat in the back of the bus, I looked back at the image of Mr. Vasquez, then drew a deep breath.

When is it going to hit? I thought. The remorse and sadness?

The bus drove off; but instead of the familiar emptiness and hollowness in my chest, another sensation filled me: It was a feeling of energy, starting at my head, then beaming and overflowing to every part of me – Till I was immersed and enveloped in what felt like a cocoon of energy, fuller and more expansive than anything I'd ever experienced before.

I've honored something, I thought. I've shown some great respect. It relates to something simple, yet wonderful, and reaches to the core of my being.

The bus drove on into the night and the feeling continued to radiate. Resting, my body conformed to the contours of the seat, as I breathed easy and felt the onset of sleep.

I want to hang on to this feeling, but I have to trust that tomorrow is another day. This connection I feel – a connection to a higher source. It radiates from my head upwards – God is love...

"I'm happy for you, Miker," Liz told me, spiritedly. "You've captured the essence of what love is all about. It has to do with little things, like compassion and respect, concern and attentiveness.

"It really isn't about me or Bubbles. It's about you – You realizing your ability to love.

"Wanting to be with Bubbles was about a desire to sooth your childhood wounds – a Band-Aid meant to control the bleeding. You can do it on your own now.

"But you still have a lot of healing to do..."

EPILOGUE

The paradigm case for healing is not to be sought in our hospitals and clinics; it is seen in those instances where faith, prayer, touch and word from a caring community have been the major factors in the healing process…
~ Paulos Mar Gregorios, Healing – A Holistic Approach

The events described here ultimately came to define my practice of Bioenergy and Chi Gong, now referred to as BioenerQi.

Recently, I led the University of California-Davis Long COVID-Qi Gong study, in which BioEnerQi was found to impart statistically significant improvements to long COVID sufferers for problems of fatigue, pain, activity tolerance, shortness of breath, anxiety and depression, as well as near statistically significant improvements for problems of sleep and cognitive function.

When I told Bubbles about the study, she responded, "I am not surprised at your results. You have been relentless in your work – work that cannot be found elsewhere – with commitment to and respect for people, willingness to share your energy with the goal of healing, guided by love and tenderness."

At the time she shared these words, I thought she might as well have been talking about herself.

Why was Bubbles so important to my BioEnerQi efforts and, really, everything I did and do? I think it's because she took it upon herself to open the door to my healing, get me beyond myself and my traumas, and help me realize my true spirit.

That's what love can do…

ABOUT THE AUTHOR

Michael Yanuck MD PhD is a physician-scientist, his groundbreaking research at the National Institutes of Health was the basis for a FDA-approved vaccine for cancer that ushered in the science of Neoantigen Tumor Immunotherapy now used to fight cancer worldwide. Following a traumatic leg injury, he trained in bioenergy and introduced energy medicine approaches at the National Institute of Complementary and Integrative Health. Now, a Professor of Medicine, he cares for the homeless, leads efforts to combat the opioid crisis and advances integrative therapies for the treatment of pain.

Made in the USA
Monee, IL
04 November 2026

67376414R00066